The Science of Creation: New Quantum Field Theory of Spaces with Octonion Cosmology Consequences

Stephen Blaha Ph. D.
Blaha Research

Quantum Field Theory of Particles Containing Child Spaces
Perturbation Theory
Fermion-Antifermion Annihilation to Scalar Space Particles
Symmetry Splitting vs. Breakdown
Dressing of Spaces with Energy and Matter

Pingree-Hill Publishing
MMXXI

Rev. 00/00/01   July 24, 2021

To Margaret

Some Other Books by Stephen Blaha

All the Megaverse! Starships Exploring the Endless Universes of the Cosmos using the Baryonic Force (Blaha Research, Auburn, NH, 2014)

SuperCivilizations: Civilizations as Superorganisms (McMann-Fisher Publishing, Auburn, NH, 2010)

All the Universe! Faster Than Light Tachyon Quark Starships & Particle Accelerators with the LHC as a Prototype Starship Drive Scientific Edition (Pingree-Hill Publishing, Auburn, NH, 2011).

Unification of God Theory and Unified SuperStandard Model THIRD EDITION (Pingree Hill Publishing, Auburn, NH, 2018).

The Exact QED Calculation of the Fine Structure Constant Implies ALL 4D Universes have the Same Physics/Life Prospects (Pingree Hill Publishing, Auburn, NH, 2019).

Unified SuperStandard Theory and the SuperUniverse Model: The Foundation of Science (Pingree Hill Publishing, Auburn, NH, 2018).

Quaternion Unified SuperStandard Theory (The QUeST) and Megaverse Octonion SuperStandard Theory (MOST) (Pingree Hill Publishing, Auburn, NH, 2020).

Unified SuperStandard Theories for Quaternion Universes & The Octonion Megaverse (Pingree Hill Publishing, Auburn, NH, 2020).

The Essence of Eternity: Quaternion & Octonion SuperStandard Theories (Pingree Hill Publishing, Auburn, NH, 2020).

A Very Conscious Universe (Pingree Hill Publishing, Auburn, NH, 2020).

From Octonion Cosmology to the Unified SuperStandard Theory of Particles (Pingree Hill Publishing, Auburn, NH, 2020).

Beyond Octonion Cosmology (Pingree Hill Publishing, Auburn, NH, 2021).

Available on Amazon.com, bn.com Amazon.co.uk and other international web sites as well as at better bookstores (through Ingram Distributors).

CONTENTS

FIGURES and TABLES

Introduction

This book develops a new quantum Field Theory of Particle Spaces and shows that it leads directly to Octonion Cosmology. It begins by defining fermion and boson space fields for an arbitrary space-time dimension. It then proceeds to develop perturbation theory and to calculate the S-matrix element for the annihilation of a fermion-antifermion pair with the production of a scalar boson. The scalar boson has an internal array that maps to the dimension array of a space using a unitary group. Subsequently the dimension array of the boson determines the pattern of internal symmetries of each space of Octonion Cosmology including internal symmetry splitting.

This process leads to the author's NEWQUeST, NEWUTMOST, and NEWUST. It provides a complete derivation of Octonion Cosmology with the resulting features of elementary particles.

The author's Theory of Everything is thereby reduced to the Quantum Field Theory of Spaces. It has a clear, direct line of development based on the author's papers in the 1970's, and books in the Twenty-First Century. The author suggests that Quantum Field Theory is the *lingua franca* of Nature.

1. Quantum Field Theory of Fermion Space Particles

Quantum Field Theory[1] has had a long history of success in dealing with elementary particles. Its primary motivation is its ability to well describe the discrete nature of particles in a quantum framework as opposed to continuous matter.

Universes (and Megaverses, and so on) which have a multidimensional character, must be handled in a different manner—they have a particulate nature—but they also have additional structure requiring a more general formulation.[2] In this book we begin by creating a Quantum Field Theory for instances of spaces containing particles and energy in a space-time together with a set of internal symmetries that evolves according to physical law.

The instances will be particles having spaces with internal dimensions. The instances will have particle features and support interactions based on lagrangian formulations. Thus the instances of the theory will exist within one space-time and yet the instances will have internal space-times of lower dimension that governs evolution within an instance. See Fig. 1.1. Space instances are more complex because of the internal evolution within the instances.

This *modus operandi* supports the generation of spaces and instances within Octonion Cosmology as illustrated by Fig. 1.2.

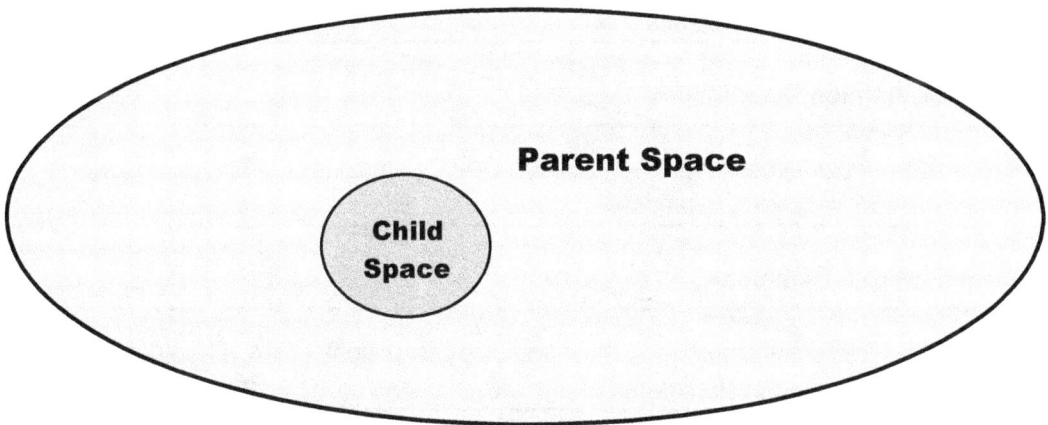

Figure 1.1. A space containing an instance that has an internal space of lower dimension. The space is itself within an instance of a higher space.

---

[1] Bjorken (1964) and (1965) are classic works in QFT. We generally follow their notation in this book and use the metric $g^{00} = +1$ with $g^{11} = g^{22} = g^{33} = -1$.

[2] We must distinguish between a space which is a multidimensional entity consisting of a space-time plus a set of internal symmetries, and an instance of the space that consists of mass and energy within the space-time governed by the internal symmetries and physical law. We shall use the word "space" and an instance interchangeably with the conext determining whether we are dealing with a space or an instance of the space. This procedure will avoid cumbersone verbiage while avoiding possible confusion.

God-Space – Space 0: Complex Octonion Octonion Octonion

1,048,576 dimensions $2^{20/2} = 1024$ rows/columns	18 Space-time Dimensions	512 512-spinors

Space 1: Octonion Octonion Octonion

262,144 dimensions $2^{18/2} = 512$ rows/columns	16 Space-time Dimensions	256 256-spinors

Space 2: Quarternion Octonion Octonion

65,536 dimensions $2^{16/2} = 256$ rows/columns	14 Space-time Dimensions	128 128-spinors

Space 3: Complex Octonion Octonion

16,384 dimensions $2^{14/2} = 128$ rows/columns	12 Space-time Dimensions	64 64-spinors

Space 4: Octonion Octonion

4,096 dimensions $2^{12/2} = 64$ rows/columns	10 Space-time Dimensions	32 32-spinors

Space 5: Quaternion Octonion

1,024 dimensions $2^{10/2} = 32$ rows/columns	8 Space-time Dimensions	16 16-spinors

Space 6: Complex Octonion

256 dimensions $2^{8/2} = 16$ rows/columns	4 Space-time Dimensions	4 4-spinors

Space 7: Octonion

64 dimensions $2^{6/2} = 8$ rows/columns	2 Space-time Dimensions	4 4-spinors
	(Built from four 16 dimension spaces of the 10 space spectrum)	

Figure 1.2. A sequence of fermion-antifermion annihilations generating instances in the set of Octonion Cosmology spaces.

We now turn to develop the Quantum Field Theory of Space Particles.

1.1 Free Space Particle Dirac Equation

In this section we will derive[3] the Space Particle Dirac equation using Lorentz boosts in r space-time dimensions of the wave function of a space particle at rest. We assume an r-dimension space-time with one time coordinate x^0 and $r - 1$ space components. In an r-dimension space time where r is an even integer, a spinor vector has $2^{r/2}$ coordinates, and a spinor array (which we introduce later) is a $2^{r/2} \times 2^{r/2}$ array.

We now turn to determine the form of the Dirac equation for an r-dimension space-time from Lorentz boosts.

1.1.1 Boosts in r Space-time Dimensions

The coordinates of inertial reference frames in r space-time dimensions are related by transformations of the r-dimensional inhomogeneous Lorentz group. To determine the form of its generators we consider infinitesimal "rotations" and translations. We denote translations by d^μ and homogeneous "rotations" by $\Lambda^\alpha{}_\beta$. The identity rotation is $\Lambda_{16}{}^\alpha{}_\beta = \delta^\alpha{}_\beta$. An infinitesimal rotation can be expressed in the form

$$\Lambda^\alpha{}_\beta = \delta^\alpha{}_\beta + \omega^\alpha{}_\beta \tag{1.1}$$

The defining feature of Lorentz transformations is

$$\Lambda^T G\, \Lambda = G \tag{1.2}$$

where Λ is a "Lorentz" transformation expressed in matrix form, Λ^T is the transpose of Λ, and G is the metric $\eta_{\mu\nu}$ expressed as the $r \times r$ diagonal matrix: G = diag(1, -1, -1, ..., -1).. In the case of infinitesimal rotations eq. 1.2 becomes

$$(\delta^\alpha{}_\mu + \omega^\alpha{}_\mu)\eta_{\alpha\beta}\,(\delta^\beta{}_\nu + \omega^\beta{}_\nu) = \eta_{\mu\nu} \tag{1.3}$$

Using $\eta_{\alpha\beta}$ to lower indices and $\eta^{\alpha\beta}$ to raise indices we have

$$\omega_{\beta\mu} = \eta_{\alpha\beta}\omega^\alpha{}_\mu \tag{1.4}$$

so that eq. 1.2 implies the antisymmetry of $\omega_{\mu\beta}$

$$\omega_{\mu\beta} + \omega_{\beta\mu} = 0 \tag{1.5}$$

to leading order in ω.

The infinitesimal, unitary, linear Hilbert space transformation including a translation by d_μ, is

$$U(\omega, d) = I + \tfrac{1}{2}\, i\, \omega_{\alpha\beta}J^{\alpha\beta} - id_\mu P^\mu \tag{1.6}$$

---

[3] These sections extend the discussion of universe particles of Blaha (2015a) and (2018e).

where the unitarity of U requires the $r(r-1)/2$ operators $J^{\alpha\beta}$ and r operators P^μ to be hermitian, and eq. 1.5 requires $J^{\alpha\beta}$ to be antisymmetric in α and β. After some further conventional considerations, the following r-dimensional Poincaré group generator algebra commutation relations result with the metric $\eta_{16\mu\nu}$ and its inverse $\eta_{16}^{\mu\nu}$.

$$[J^{\alpha\beta}, J^{\kappa\lambda}] = i(\eta_{16}^{\alpha\varkappa}J^{\beta\lambda} + \eta_{16}^{\lambda\alpha}J^{\kappa\beta} - \eta_{16}^{\beta\varkappa}J^{\alpha\lambda} - \eta_{16}^{\lambda\beta}J^{\kappa\alpha}) \qquad (1.7)$$

$$[P^\alpha, J^{\kappa\lambda}] = i(\eta_{16}^{\alpha\lambda}P^\kappa - \eta_{16}^{\alpha\varkappa}P^\lambda) \qquad (1.8)$$

$$[P^\alpha, P^\kappa] = 0 \qquad (1.9)$$

The momentum operators are

$$\mathbf{P} = (P^0, P^1, \dots, P^{r-1}) \qquad (1.10)$$

The boost $(r-1)$-vector is

$$\mathbf{K} = (J^{10}, J^{20}, \dots, J^{(r-1)0}) \qquad (1.11)$$

The angular momentum $(r-1)$-vector is

$$\mathbf{J} = (J^{(r-2)(r-1)}, J^{(r-1)1}, J^{12}, J^{23}, J^{34}, \dots, J^{(r-3)(r-2)}) \qquad (1.12)$$

The form of an inhomogeneous r-dimensional Lorentz, Hilbert space transformation is

$$U(\mathbf{v}, \boldsymbol{\theta}, \mathbf{d}) = \exp[i\mathbf{a}\cdot\mathbf{K} + i\boldsymbol{\theta}\cdot\mathbf{J} - i\mathbf{d}\cdot\mathbf{P}] \qquad (1.13)$$

where $\mathbf{a}$, $\boldsymbol{\theta}$ and $\mathbf{d}$ are complex.

A homogeneous Lorentz transformation $S(\mathbf{v}) = U(\mathbf{v}, \boldsymbol{\theta} = 0, \mathbf{d} = 0)$ on r-space-time dimension Dirac matrices (of size $2^{r/2} \times 2^{r/2}$ rows and columns) is[4]

$$S^{-1}(v)\gamma^\nu S(v) = \Lambda^\nu_{\ \mu}(v)\gamma^\mu \qquad (1.14)$$

where $\Lambda^\nu_{\ \mu}(v)$ is the corresponding Lorentz coordinate transformation, and the inverse satisfies

$$S^{-1}(v) = \gamma^0 S^\dagger(v)\gamma^0 \qquad (1.15)$$

where † signifies Hermitian conjugate.

---

[4] Blaha (2007b) gives the explicit form of s for four dimensions.

1.1.2 Generation of Fermion Space Particle Wave Function

We now derive the fermion space particle wave function using a Lorentz boost. We specify a generic positive energy plane wave solution of the Dirac equation for a space particle at rest with mass m as

$$\psi(x) = e^{-imt}W(0) \tag{1.16}$$

with $W(0)$ being a $2^{r/2} \times 2^{r/2}$ matrix (which is a generalization of the $2^{r/2}$ component spinor column vector for a normal fermion.)

$\psi(x)$ satisfies the momentum space Dirac equation for a space particle at rest:

$$(m\gamma^0 - m)e^{-imt}W(0) = 0 \tag{1.17}$$

$W(0)$ has the form $W(0) = \text{diag}(1, 1, 1, \ldots, 1)$. If we apply $S(v)$ we find

$$0 = S(v)(m\gamma^0 - m)e^{-imt}W(0) = [mS(v)\gamma^0 S^{-1}(v) - m]S(v)W(0)$$

We now require the Lorentz transformation to satisfy

$$mS(v)\gamma^0 S^{-1}(v) = g_{\mu\nu}p^\mu\gamma^\nu = \not{p} \tag{1.18}$$

where $p^0 = (p^2 + m^2)^{1/2}$, $\mathbf{p} = \gamma m\mathbf{v}$, and $p = |\mathbf{p}|$ in r-dimension space-time. In addition we define

$$W(p) = S(v)W(0) \tag{1.19}$$

Therefore the free space particle Dirac equation in momentum space has the form:

$$(\not{p} - m)e^{-ip \cdot x}W(p) = 0 \tag{1.20}$$

where the exponential factor, "mt", is also boosted to p·x. Eq. 1.20 implies the free, coordinate space Dirac equation:

$$(i\gamma^\mu \partial/\partial x^\mu - m)\psi(x) = 0 \tag{1.21}$$

1.1.3 Form of the Positive Frequency W Matrix

Eq. 1.18 determines the form of $S(\mathbf{v})$ in terms of p:

$$S(v) = S(p) = (2m(m + p^0))^{-1/2}\gamma^0(\not{p} + m\gamma^0) \tag{1.22}$$

and

$$S^{-1}(v) = S^{-1}(p) = (2m(m + p^0))^{-1/2}\gamma^0(\not{p} + m\gamma^0)\gamma^0 \tag{1.23}$$

where p^0 is the energy. Note eqs. 1.22 and 1.23 implement eqs. 1.15 and 1.18.

1.1.3 Form of the Negative Frequency (Hole) W Matrix

The form of the Dirac equation for negative frequency is

$$(\not{p} + m)e^{-ip \cdot x}W'(p) = 0 \tag{1.24}$$

Eq. 1.24 for the negative frequency Dirac equation requires a different Lorentz transformation $S'(p)$. The equation

$$mS'(p)\gamma^0 S'^{-1}(p) = -g_{\mu\nu}p^\mu\gamma^\nu = -\not{p} \tag{1.25}$$

determines the form of $S'(\mathbf{v})$ in terms of p:

$$S'(p) = (2m(m - p^0))^{-\frac{1}{2}}\gamma^0(-\not{p} + m\gamma^0) \tag{1.26}$$

and

$$S'^{-1}(p) = (2m(m - p^0))^{-\frac{1}{2}}\gamma^0(-\not{p} + m\gamma^0)\gamma^0 \tag{1.27}$$

where p^0 is the energy. Consequently

$$W'(p) = S'(p)W(0) \tag{1.28}$$

Therefore the negative frequency free space particle Dirac equation in momentum space has the form:

$$(\not{p} + m)e^{-ip \cdot x}W'(p) = 0 \tag{1.29}$$

1.2 Wave Particle Expansion

Having determined the form of $S(v)$ and $S'(p)$, and thus $W(p)$ and $W'(p)$ we can specify the form of an r space-time dimension fermion *space* wave function with one time dimension and r – 1 spatial dimensions. We will use the form of the Lorentz boost eq. 1.14 since it determines the spinor factors in the fermion wave function. We will also specify the form of a scalar boson for a space particle using the form of the Lorentz boost. (Fermion-antifermion annihilation into the scalar boson will show that the scalar boson wave function "uses" the fermion spinor factors as an input.) The spinor arrays U and V have their size determined by r, which we call the *parent* space space-time dimension. The space-time dimension q is the *child* space space-time dimension.

The space fermion wave function for an r space-time dimension space fermion (with a q-dimension inner space-time) is

$$\Psi_a(r, q, x) = \int d^{r-1}p(2\pi)^{-(r-1)}(m/p^0)^{\frac{1}{2}} \{exp^{-ip \cdot x} U_{a\beta}(r, q, p)b_\beta(p) +$$

$$+ exp^{ip \cdot x} V_{a\beta}(r, q, p)d^\dagger_\beta(p)\} \tag{1.30}$$

and its Hermitian conjugate is

$$\Psi^\dagger_a(r, q, x) = \int d^{r-1}p(2\pi)^{-(r-1)}(m/p^0)^{\frac{1}{2}} \{ exp^{ip\cdot x}\, b^\dagger_\beta(p)U^\dagger_{\beta a}(r, q, p) +$$

$$+ exp^{-ip\cdot x}\, d_\beta(p)V^\dagger_{\beta a}(r, q, p) \} \quad (1.31)$$

One may visualize the form of the spinor matrix U as a matrix with $2^{r/2}$ spinor columns, in which each column is a spinor with $2^{r/2}$ components. (Fig. 1.3)

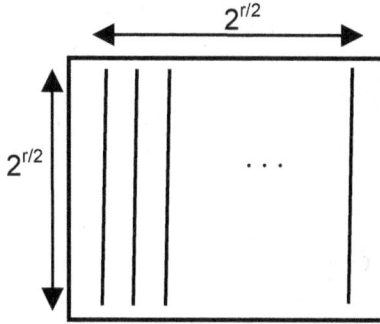

Figure 1.3. The U array with $2^{r/2}$ columns and $2^{r/2}$ rows. The V array has the same form.

Eq. 1.30 expresses Ψ as a sum over spins with a summation over β. For each value of β there is a creation/annihilation operator (b_β and $d_\beta^\dagger$) and a spinor vector $U_{\alpha\beta}(r, q, p)$/ $V_{\alpha\beta}(r, q, p)$ with indices labeled by α *as rows* within the spinor array. For each column spinor, α labels the components within it.

We define the spinor arrays ($2^{r/2} \times 2^{r/2}$) with

$$U(r, q, p, m) = ((m + p^0)/2m)^{\frac{1}{2}}\gamma^0 S(p)$$
$$= (\not{p} + m)/2m \quad (1.32)$$

$$V(r, q, p, m) = ((m - p^0)/2m)^{\frac{1}{2}}\gamma^0 S'(p)$$
$$= (m - \not{p})/2m \quad (1.33)$$

Note that U and V are projections satisfying

$$U(r, q, p, m)^2 = U(r, q, p, m) \quad (1.34)$$
$$V(r, q, p, m)^2 = V(r, q, p, m)$$

$$U(r, q, p, m)V(r, q, p, m) = V(r, q, p, m)U(r, q, p, m) = 0 \quad (1.35)$$

Eq. 1.30 cleanly separates the positive and negative frequency parts of $\Psi(r, q, x)$.
The spinor arrays are normalized column by column:

1. The columns of U form an orthonormal set of $2^{r/2}$-vectors.
2. The columns of V form an orthonormal set of $2^{r/2}$-vectors.

$$[\gamma^0 U(r, q, p, m)]^\dagger{}_{\alpha\beta} U(r, q, p, m)_{\beta\sigma} = [\gamma^0 U(r, q, p, m)]_{\alpha\sigma}$$
$$[\gamma^0 V(r, q, p, m)]^\dagger{}_{\alpha\beta} V(r, q, p, m)_{\beta\sigma} = [\gamma^0 V(r, q, p, m)]_{\alpha\sigma}$$
$$U(r, q, p, m)\gamma^0 U(r, q, p, m)^\dagger = U(r, q, p, m)\,\gamma^0 \qquad (1.36)$$
$$V(r, q, p, m)\gamma^0 V(r, q, p, m)^\dagger = V(r, q, p, m)\,\gamma^0$$

and

$$U(r, q, p, m)_{\alpha\beta} U(r, q, p, m)^\dagger{}_{\beta\sigma} = (E/m)[U(r, q, p, m)\gamma^0]_{\alpha\sigma}$$
$$V(r, q, p, m)_{\alpha\beta} V(r, q, p, m)^\dagger{}_{\beta\sigma} = -(E/m)[V(r, q, p, m)\gamma^0]_{\alpha\sigma} \qquad (1.37)$$

with $p^0 = E$. Note:

$$\gamma^0 U(r, q, p, m)^\dagger = U(r, q, p, m)\,\gamma^0 \qquad (1.37a)$$

$$\gamma^0 V(r, q, p, m)^\dagger = V(r, q, p, m)\,\gamma^0 \qquad (1.37b)$$

The creation and annihilation operators satisfy fermion anticommutation relations:

$$\{b_\beta(p), b^\dagger{}_\sigma(p')\} = \delta_{\beta\sigma}\, \delta^{r-1}(p - p') \qquad (1.38)$$
$$\{b_\beta(p), b_\sigma(p')\} = 0$$
$$\{b^\dagger{}_\beta(p), b^\dagger{}_\sigma(p')\} = 0$$

$$\{d_\beta(p), d^\dagger{}_\sigma(p')\} = \delta_{\beta\sigma}\, \delta^{r-1}(p - p') \qquad (1.39)$$
$$\{d_\beta(p), d_\sigma(p')\} = 0$$
$$\{d^\dagger{}_\beta(p), d^\dagger{}_\sigma(p')\} = 0$$

Note β and ρ are spin labels for the $2^{r/2}$ spin states. The spin of the fermion is

$$s = 2^{r/2 - 1} - \tfrac{1}{2} \qquad (1.40)$$

In four space-time dimensions the *space* fermion has spin 3/2. The fermion spin in various spaces may be high. This is not a problem remembering that higher spin resonances appear on Regge trajectories for example. They are also not a problem in perturbation theory if one uses our Two Tier form of Quantum Field Theory that eliminates divergences in perturbation theory for any number of space-time dimensions. See Blaha (2005a).

1.3 Fermion Space Particle Interpretation

Fermion space particles have the space-time dependence specified by their Dirac equations. Their spinor behavior is embodied in the $U(r, q, p)$ and $U(r, q, p)$ arrays. In chapter 2 we will show the spinor arrays have a dual role: they specify fermion spin and they furnish the dimension arrays of the Octonion Spectrum of Spaces. In their second role they become a space, which itself has a space-time symmetry as well as internal symmetries.

1.4 Feynman Propagator

Fermion space particle wave functions have the anticommutation relation:

$$\{\Psi_a(r, q, t, \mathbf{x}), \Psi^\dagger_b(r, q, t, \mathbf{x}')\} = \iint d^{r-1}p\ d^{r-1}p' \delta^{r-1}(\mathbf{p} - \mathbf{p}')\{U_{a\alpha}(p)U^\dagger_{\alpha b}(p')\ e^{ip\cdot(x-x')} +$$

$$+ V_{a\alpha}(p)V^\dagger_{\alpha b}(p')\ e^{-ip\cdot(x-x')}\}(m^2/EE')^{1/2}\ (2\pi)^{-3}$$

$$= \gamma^0_{ab}\ \delta^{r-1}(\mathbf{x} - \mathbf{x}') \tag{1.41}$$

with other anticommutators zero, and with $p^0 = E$ and using eqs. 1.37 with the r, q, and m arguments surpressed in U and V.

The Feynman propagator is:

$$(S_F(x' - x)\ \gamma^0)_{ab} = -i<0|\Psi_a(r, q, x')\gamma^0\Psi^\dagger_b(r, q, x)|0>\theta(t' - t) +$$
$$+ i <0| \Psi^\dagger_b(r, q, x)\ \gamma^0\Psi_a(r, q, x')|0>\theta(t - t')$$

$$= \iint d^{r-1}p\ d^{r-1}p' \delta^{r-1}(\mathbf{p} - \mathbf{p}')\{-i[U_{a\alpha}(p)\ \gamma^0 U^\dagger_{\alpha b}(p')e^{ip\cdot(x-x')}]\theta(t' - t) +$$

$$+ i\ [V_{a\alpha}(p)\ \gamma^0 V^\dagger_{\alpha b}(p')\ e^{-ip\cdot(x-x')}]\theta(t - t')\}(m^2/EE')^{1/2}(2\pi)^{-(r-1)}$$

$$= \int d^{r-1}p\ \{-ie^{ip\cdot(x-x')}\ [U(p)\gamma^0]_{ab}\ \theta(t' - t) +$$

$$+ ie^{-ip\cdot(x-x')}[V(p)\gamma^0]_{ab}\ \theta(t - t')\}(2\pi)^{-(r-1)}\ (m/E)$$

using eq. 1.36

$$= -i \int d^{r-1}p\ (m/E)\{e^{ip\cdot(x-x')}\ [U(p)\gamma^0]_{ab}\ \theta(t' - t) +$$

$$+e^{-ip\cdot(x-x')}[V(p)\ \gamma^0]_{ab}\ \theta(t - t')\}(2\pi)^{-(r-1)}$$

Or

$$S_F(x' - x) = -i\int d^{(r-1)}p\{e^{-ip\cdot(x'-x)}U(p)_{ab}\ \theta(t' - t) + e^{ip\cdot(x'-x)}V(p)_{ab}\ \theta(t - t')\}(2\pi)^{-(r-1)} \tag{1.42}$$

$$= \int d^{(r-1)}p\ e^{-ip\cdot(x'-x)}\ (2\pi)^{-r}\ (\not{p} + m)/(p^2 - m^2 + i\varepsilon)$$

Thus the Feynman propagator for space fermions is the same as that of "normal" fermions for r dimension space-time.

2. Scalar Space Particle Quantum Field Theory

Second quantization of particle wave functions uses Bose-Einstein quantization for bosons and Fermi-Dirac quantization for Fermions. Their origin is based in observations of particle behavior and the Pauli Exclusion Principle.

The Fermi-Dirac fermion quantization procedure is not required theoretically.[5,6] In this chapter we will define a scalar space particle field. We will find it convenient to use Bose-Einstein quantization in view of its scalar nature.[7] But we will formulate it internally as having the spinor form of a space fermion in order to reach the goal of having fermion-antifermion annihilation produce a scalar[8] space particle containing an internal symmetry array. We will see that Bose-Einstein leads directly to the spaces of Octonion Cosmology.

2.1 Lagrangian and Dirac Equation

We define the Lagrangian for the scalar space particle interacting with the fermion space particle to be:

$$\mathcal{L} = \overline{\psi}(i\partial_\mu \gamma^\mu - g\Phi - m)\psi + \overline{\Phi}(i\partial_\mu \gamma^\mu - m')\,\Phi \ \ldots \quad (2.1)$$

The free scalar space particle Dirac equation is:

$$(i\partial_\mu\gamma^\mu - m')\,\Phi(r, q, x) = 0 \qquad (2.2)$$

2.2 Scalar Space Particle Wave Function

Emulating the fermion wave function of the previous chapter, we choose the scalar space particle wave function for an r space-time dimension space boson (with a q-dimension inner space-time) to be

---

[5] The only possible source of issues is a potential cascade into negative energy states. But the negative energy state issue is obviated by the characterization of negative energy states as positive energy states moving backward in time.

[6] Knowing that some will object to our quantization procedure the author acknowledges that he is *solely* responsible for his theoretical results. He has not received any feedback, suggestions, or comments. Contrary to some rumors the author has had no academic affiliation in the past twenty years, and only an honorary affiliation in the preceding 20 years. In the past 10 years the author has not had contact with other physicists. In the past 20 years there were a few very brief contacts of a merely social nature so that the author could pursue his original line of research unencumbered by preconceptions. Blaha Research is an independent, privately-funded research institute founded in 2000 with the purpose of advancing human knowledge expeditiously.

[7] There is a resemblance between Bose-Einstein even spin and Fermi-Dirac half integer spin phenomena. This is illustrated by the author's discovery (S. Blaha, Phys. Rev. Lett., **36**, 874 (1976)) of Fermi Superfluid He[3] quantized vortices, which resemble vortices in Bose superfluid helium.

[8] We take "scalar" to mean having Bose-Einstein statistics that allow any number of scalar particles with the same quantum numbers. Scalar particles do not obey the Pauli Exclusion Principle.

$$\Phi_a(r, q, x) = \int d^{r-1}p(2\pi)^{-(r-1)}(m'/p^0)^{\frac{1}{2}} \{exp^{-ip \cdot x} U_{a\beta}(r, q, p, m')a_\beta(p) +$$

$$+ exp^{ip \cdot x} V_{a\beta}(r, q, p, m')c^{\dagger}_\beta(p)\} \quad (2.3)$$

and its Hermitian conjugate

$$\Phi^{\dagger}_a(r, q, x) = \int d^{r-1}p(2\pi)^{-(r-1)}(m'/p^0)^{\frac{1}{2}} \{exp^{ip \cdot x} a^{\dagger}_\beta(p)U^{\dagger}_{\beta a}(r, q, p, m') +$$

$$+ exp^{-ip \cdot x} c_\beta(p)V^{\dagger}_{\beta a}(r, q, p, m') \} \quad (2.4)$$

where m' is the scalar particle mass, and U and V are the same as in fermion case. The spinor arrays, U and V, have $2^{r/2}$ columns and $2^{r/2}$ rows.. The spinor arrays U and V have their size determined by r, which we call the *parent* space space-time dimension. The space-time dimension q is the *child* space space-time dimension.

The creation/annihilation operators satisfy the commutation relations:

$$[a_{jk}(p), a^{\dagger}_{j'k'}(p')] = \delta_{\beta\sigma j'k' j'k'} \delta^{r-1}(p - p') \quad (2.5)$$
$$[a_\beta(p), a_\sigma(p')] = 0$$
$$[a^{\dagger}_\beta(p), a^{\dagger}_\sigma(p')] = 0$$

plus similar commutation relations for $c_\beta(p)$ and $c^{\dagger}_\sigma(p')$.

There are $2^{r/2}$ operators $a_\beta(p)$ and $2^{r/2}$ c_β. They can be viewed as corresponding to $2^{r/2}$ particles of an internal symmetry. We take this view in chapter 4 when we introduce a $U(2^{r/2})$ symmetry. For the moment we use them to calculate the commutator, and Feynman propsgator, of the Φ field.

Scalar space particle wave functions have the commutation relations:

$$[\Phi_a(r, q, t, \mathbf{x}), \Phi^{\dagger}_b(r, q, t, \mathbf{x'})] = \iint d^{r-1}p \, d^{r-1}p' \delta^{r-1}(\mathbf{p} - \mathbf{p'})\{U_{a\alpha}(p)U^{\dagger}_{\alpha b}(p') \, e^{ip \cdot (x - x')} +$$

$$+ V_{a\alpha}(p)V^{\dagger}_{\alpha b}(p') \, e^{-ip \cdot (x - x')}\}(m^2/EE')^{\frac{1}{2}} (2\pi)^{-3}$$

$$= \gamma^0_{ab} \delta^{r-1}(\mathbf{x} - \mathbf{x'}) \quad (2.6)$$

with $p^0 = E$ and using eqs. 1.37 with the r, q, and m arguments in U and V not displayed.

The Feynman propagator is:

$$(S_F(x' - x) \gamma^0)_{ab} = -i<0|\Phi_a(r, q, x')\gamma^0\Phi^{\dagger}_b (r, q, x)|0>\theta(t' - t) +$$
$$+ i <0| \Phi^{\dagger}_b (r, q, x)\gamma^0\Phi_a(r, q, x')|0>\theta(t - t')$$

$$= \iint d^{r-1}p \, d^{r-1}p' \delta^{r-1}(\mathbf{p} - \mathbf{p'})\{-i[U_{a\alpha}(p) \gamma^0 U^{\dagger}_{\alpha b}(p')e^{ip \cdot (x - x')}]\theta(t' - t) +$$

$$+ i [V_{a\alpha}(p) \gamma^0 V^\dagger_{\alpha b}(p') e^{-ip\cdot(x-x')}]\theta(t-t')\}(m'^2/EE')^{\frac{1}{2}}(2\pi)^{-(r-1)}$$

$$= \int d^{r-1}p \{-ie^{ip\cdot(x-x')} [U(p)\gamma^0]_{ab} \theta(t'-t) +$$

$$+ ie^{-ip\cdot(x-x')}[V(p)\gamma^0]_{ab} \theta(t-t')\}(2\pi)^{-(r-1)} (m'/E)$$

using eq. 1.36.

$$= -i \int d^{r-1}p (m'/E)\{e^{ip\cdot(x-x')} [U(p)\gamma^0]_{ab} \theta(t'-t) +$$

$$+e^{-ip\cdot(x-x')}[V(p) \gamma^0]_{ab} \theta(t-t')\}(2\pi)^{-(r-1)}$$

Or

$$S_F(x'-x) = -i\int d^{(r-1)}p\{e^{-ip\cdot(x'-x)}U(p)_{ab} \theta(t'-t) + e^{ip\cdot(x'-x)}V(p)_{ab} \theta(t-t')\}(2\pi)^{-(r-1)}$$

$$(2.7)$$

$$= \int d^{(r-1)}p\, e^{-ip\cdot(x'-x)} (2\pi)^{-r} (\not{p} + m')/(p^2 - m'^2 + i\varepsilon)$$

Thus the form of the Feynman propagator for space scalars is the same as that for space fermions for r dimension space-time.

A scalar space particle has an r dimension space-time "skin" with r-momentum p, and contains a space which includes a q dimension internal space-time.

3. Space Particle Perturbation Theory

The general form of perturbation theory is the same except for self-evident changes in such quantities as the number of dimensions. Bjorken (1965) has the traditional approach. However higher dimension calculations have divergences that cannot be eliminated by standard 4-dimension renormalization. The author's Two Tier approach,[9] which defines quantum coordinates, is based on

$$X^\mu(y) = y^\mu + i\, Y^\mu(y)/M_c^2 \qquad (3.1)$$

where $Y^\mu(y)$ is a QED-like vector boson field, M_c is an extremely large mass, and the y^μ are the underlying coordinates. The two tiers of coordinates are the quantum coordinates $X^\mu(y)$ and the c-number coordinates y^μ. Two Tier coordinates eliminate *all* divergences (including fermion triangle divergences) in *all* dimensions.

Thus a *completely finite* perturbation theory is supported for all space-time dimensions r. See Blaha (2005a) for the Two Tier form of perturbation theory. It is similar to standard perturbation theory in most respects. *Its "low energy" limit for calculations yields conventional quantum field theory results.*

3.1 Fermion-Antifermion Annihilation into Scalar Space Particle

In this section we consider the creation of a scalar space particle by offshell[10] fermion-antifermion annihilation. We will use the model Lagrangian density:

$$\mathcal{L} = \overline{\psi}(i\partial_\mu \gamma^\mu - g\Phi - m)\psi + \overline{\Phi}(i\partial_\mu \gamma^\mu - m')\,\Phi \ \dots \qquad (3.2)$$

where Φ is a space boson field array whose internal symmetry dimensions have the same number of rows and columns as the Dirac γ matrices. There is an interaction between the fermion spin and the internal symmetries of Φ that is analogous to the spin-orbit interaction in atomic physics.

The resulting lowest order annihilation process appears in Fig. 3.1. The S-matrix element is

$$S_{fi}^{\ pair} = N\,\delta^r(p_+ + p_- - p_s)\, V^\dagger(r, q, p_+, m)\gamma^0\, U(r, q, p_+ + p_-, m')U(r, q, p_-, m)$$
$$(3.3)$$

where N is a constant. $S_{fi}^{\ pair}$ is an array of spin components. Individual S-matrix elements may be obtained by selecting specific in and out spins.

---

[9] See Blaha (2005a)

[10] Offshell is required for energy-momentum conservation. An onshell process would require two scalar spaces to be created.

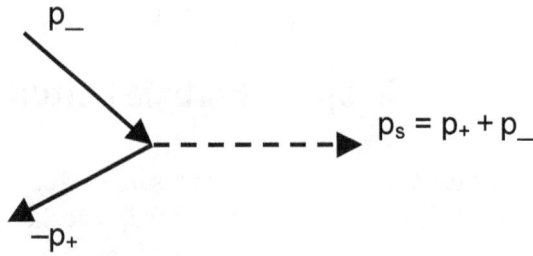

Figure 3.1. Diagram for the offshell annihilation of a space fermion – space antifermion pair to produce a space scalar particle.

Upon substituting for U and V we obtain;

$$S_{fi}^{pair} = N \, \delta^r(p_+ + p_- - p_s) \, (m - \not{p}_+)^\dagger \gamma^0 \, (\not{p}_+ + \not{p}_- + m')(\not{p}_- + m)/(8m^2m')$$

$$= N \, \delta^r(p_+ + p_- - p_s) \, \gamma^0(m - \not{p}_+)(\not{p}_- + m)/(8mm') \qquad (3.4)$$

$$= N \, \delta^r(p_+ + p_- - p_s) \, \gamma^0 V(r, q, \not{p}_+) \, U(r, q, \not{p}_-)/2$$

$$= N \, \delta^r(p_+ + p_- - p_s) \, V^\dagger(r, q, \not{p}_+)\gamma^0 U(r, q, \not{p}_-)/2$$

$$= N \, \delta^r(p_+ + p_- - p_s) \, \overline{V}(r, q, \not{p}_+)U(r, q, \not{p}_-)/2 \qquad (3.5)$$

$$= N \, \delta^r(p_+ + p_- - p_s) \, ((m - p_+^0)(m + p_-^0))^{\frac{1}{2}}/(2m) \, S'(p_+)\gamma^0 S(p_-) \qquad (3.6)$$

Comments:

1. S_{fi}^{pair} is independent of the Φ mass m'.

2. If $p_- = p_s$ then $S_{fi}^{pair} = 0$ by eq. 1.35.

3. Cayley-Dickson numbers beyond octonions have non-trivial zero divisors where $n_1 \cdot n_2 = 0$, and n_1 and n_2 are non-zero Cayley-Dickson numbers. For each type of Cayley-Dickson number there are a *finite* number of non-trivial zero divisors.

   Arrays of Cayley-Dickson numbers were considered by the author in Blaha (2021a) – (2021c). The arrays (m - $\not{p}$) and ($\not{p}$ + m) are each an array of Cayley numbers for each value of p. Together they form an *infinite* set of non-trivial zero divisors.

4. The space scalar particle internal symmetry array is generated by the fermion-antifermion annihilation. The form of the space scalar particle internal symmetry array $(p_+ + p_- + m)/2m$ follows from momentum conservation and leads to the internal symmetries of the octonion space spectrum. See chapter 7 for details.

4. Symmetry Group of Φ and Ψ

The space particle wave functions have two symmetries based on momentum conservation, and on a symmetry group of Dirac matrices. The second symmetry is the more important.

4.1 Symmetry Group Related to Momentum Conservation

Examining the calculation of a scalar space particle from fermion-antifermion annihilation, eq. 3.3, we see the fermion and antifermion momenta combine to give the momentum of the scalar space particle:

$$V \sim (\not p_+) \quad \text{"+"} \quad U \sim (\not p_-) \quad \rightarrow \quad U \sim (\not p_+ + \not p_-) \tag{4.1}$$

because of momentum conservation at the interaction vertex. Conservation may be viewed at the level of Lorentz group boosts:

$$S(p_+) \text{ "×" } S(p_-) \rightarrow S(p_+ + \underline{p}) \tag{4.2}$$

using eq. 1.22 for Lorentz group boosts.

Momentum conservation then leads to an additive abelian symmetry of the spinor arrays. If we define the traceless spinor arrays:

$$U_t(p) = \not p \tag{4.3}$$
$$V_t(p) = \not p$$

then

$$U_t(p) + U_t(s) = U_t(p + s) \tag{4.4}$$

and similarly for $V_t(p)$.

4.2 Symmetry Group of Dirac Matrices

The Dirac γ matrices of an r-dimension space-time have a $U(2^{r/2})$ unitary symmetry:[11]

$$\gamma'^{\mu} = U\gamma^{\mu}U^{-1}$$

where $U^{-1} = U^{\dagger}$ and where γ'^{μ} is an equivalent Dirac matrix.

We now consider the implications for a scalar space particle with wave function:

$$\Phi_a(r, q, x) = \int d^{r-1}p(2\pi)^{-(r-1)}(m'/p^0)^{\frac{1}{2}} \{\exp^{-ip\cdot x} U_{\alpha\beta}(r, q, p, m')a_\beta(p) +$$

---

[11] The case of 4-dimension space-time was presented in R. H. Good, Jr., Rev. Mod. Phys **27**, 187 (1955). The generalization to r –dimension space-time is direct.

$$+ \exp^{ip \cdot x} V_{\alpha\beta}(r, q, p, m')c^{\dagger}_{\beta}(p)\} \quad (4.5)$$

A $U(2^{r/2})$ transformation, Y, on the "spinor" factors of the terms gives

$$U_{\alpha\beta}(r, q, p, m')a_{\beta}(p) \rightarrow YU(r, q, p, m')Y^{-1}Ya(p) \quad (4.6)$$
$$= U'(r, q, p, m') \, a'(p)$$

and

$$V_{\alpha\beta}(r, q, p, m')c^{\dagger}_{\beta}(p) \rightarrow YV(r, q, p, m') \, Y^{-1} \, Yc^{\dagger}(p)$$
$$= V'(r, q, p, m') \, c'(p)$$

Eqs. 4.6 shows that the "spinor" array and creation/annihilation operators support $U(2^{r/2})$ unitary transformations. Thus the symmetry group $U(2^{r/2})$ describes transformations of the internal space of Φ.

We thus have shown that Φ has an internal space of 2^r total dimensions embodied in the $U(r, q, p, m')$ array. This array is then the template for the internal space, which contains q space-time dimensions and $2^r - q$ dimensions allocated to internal symmetries.

In chapter 6 we show the small and large sectors of the $U(r, q, p, m')$ spinor array correspond directly to the splitting of the internal symmetries of octonion spaces into the various symmetry groups. This match constitutes support for the concept of fermion-antifermion annihilation generating the splitting of octonion space internal symmetry groups.

The $U(r, q, p, m')$ spinor array of Φ is a dimension array for an octonionm space.

5. Specification of Octonion Spaces

The space particles that we defined in earlier chapters support the Spectrum of Octonion Spaces that we developed in earlier books. We viewed the spectrum as originating in a cascade of fermion-antifermion annihilations as pictured in Fig. 1.2. We can express the cascade of spaces in terms of the scalar space particle wave function as

$$
\begin{array}{ll}
\mathbf{r} \quad \mathbf{q} & \\
\Phi(20, 18, x) & \quad (5.1) \\
\Phi(18, 16, x) & \\
\Phi(16, 14, x) & \\
\Phi(14, 12, x) & \\
\Phi(12, 10, x) & \\
\Phi(10, \ 8, x) & \\
\Phi(8, \ \ 6, x) & \\
\Phi(6, \ \ 4, x) & \\
\Phi(4, \ \ 2, x) & \\
\Phi(2, \ \ 0, x) & \\
\end{array}
$$

The spaces, with $q = r - 2$, have space-time dimensions arranged to yield the Cayley-Dickson number pattern in Fig. 6.1. The sequence of space arrays is listed in Fig. 5.1.

Cayley-Dickson Number n	Parent Space-time Dimension r	Total Dimension Array d_c	Child Space-time Dimensions q
10	20	1024×1024	18
9	18	512×512	16
8	16	256×256	14
7	14	128×128	12
6	12	64×64	10
5	10	32×32	8
4	8	16×16	6
3	6	8×8	4
2	4	4×4	2
1	2	2×2	0

Figure 5.1. Cayley-Dickson number n, and the corresponding tnumbers of total dimensions and of space-time dimensions. The number n = 1 corresponds to complex numbers. The number n = 10 corresponds to complex octonion octonion octonion numbers (with 1024 components).

The regularity of the list reflects the choice of space-time dimensions in eq. 5.1. The complete set of ten spaces is generated by nine nested fermion-antifermion annhilations.

Since time in our universe does not exist outside of the universe, there is no issue of the amount of time required.

The cascade can generate side branches as shown in Fig. 5.2 below from Blaha (2021b). Sibling spaces are allowed—especially since we specify Bose-Einstein quantization of Φ.

Our Quantum Field Theory of Spaces supports, and is consistent with, the Octonion Cosmology Spectrum of Spaces.

A COSMOS

SPACE

INSTANCES

God-Space – Space 0:

Space 1:

Space 2:

Space 3:

Space 4:

Space 5:

Space 6:

homeverse

Space 7:

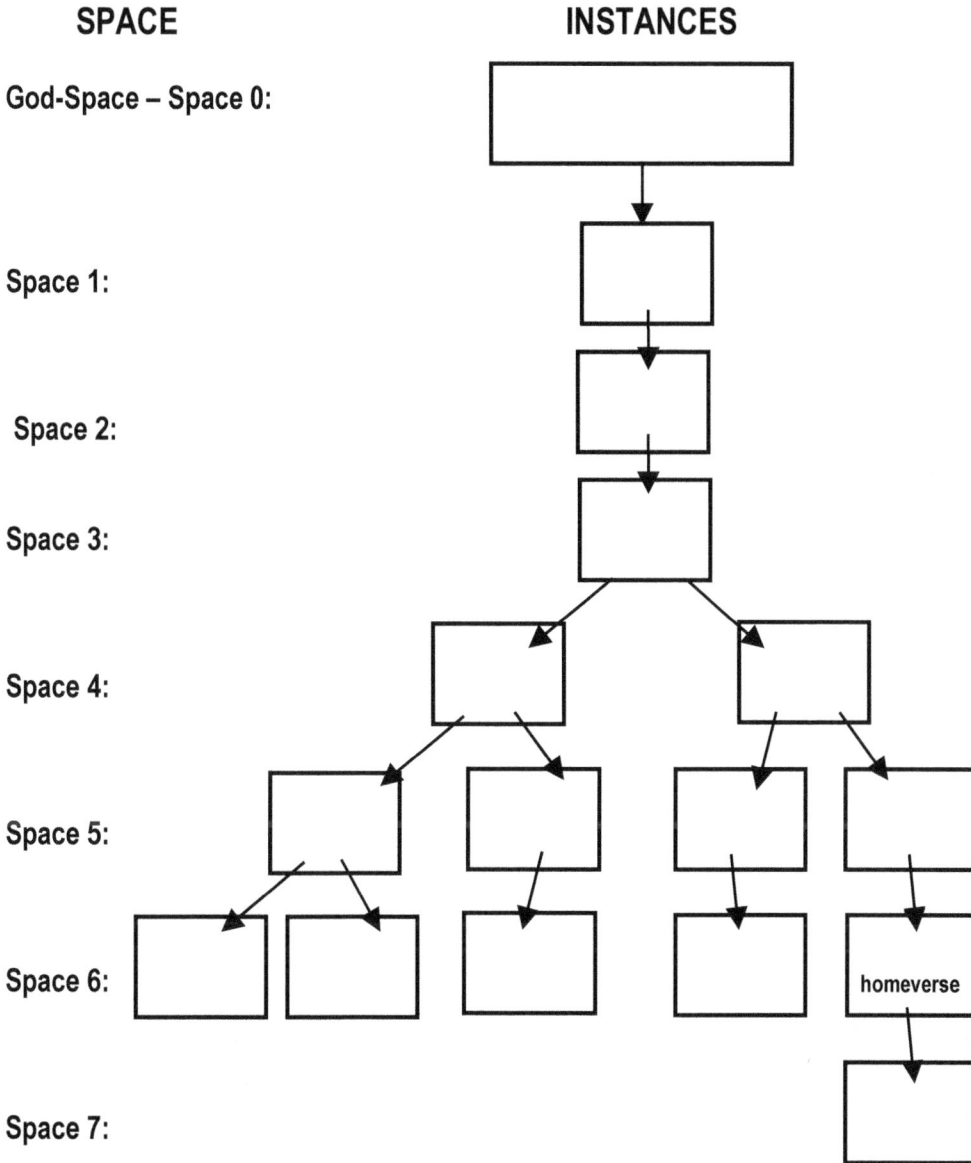

Figure 5.2. A hierarchy of instances leading from the God-Space (Cayley space 10) to "homeverse" – our designation for our universe. The homeverse is shown to contain one space 7 instance for illustration purposes. Space 7 is viewed here as combining spaces 7, 8, 9 and 10, which are part of the 10 space form of the spectrum. The homeverse has one "sibling" and three "cousin" universes. *The entire hierarchy resides in God-Space because the inheritance stems from the God-Space instance as parts of it. Other universes can be "reached" through the God-Space instance if a mode of transportation existed.*

.

6. Octonion Spaces Spectrum

The octonion spaces spectrum of Fig. 5.1 specifies ten spaces. Fig. 6.1 specifies it in more detail.

THE TEN OCTONION SPACES SPECTRUM

Octonion Space Number O_s	Cayley-Dickson Number n	Cayley Number Dimension d_c	Dimension Array Total d_d	Space-time-Dimension q	Fermion Spinor Array Total d_s	Cayley Number "Name"
0	10	1024	1024×1024	18	512×512	Complex Octonion Octonion Octonion
1	9	512	512×512	16	256×256	Octonion Octonion Octonion
2	8	256	256×256	14	128×128	Quaternion Octonion Octonion
3	7	128	128×128	12	64×64	Complex Octonion Octonion
4	6	64	64×64	10	32×32	Octonion Octonion
5	5	32	32×32	8	16×16	Quaternion Octonion
6	**4**	**16**	**16×16**	**6^{12}**	**8×8**	**Complex Octonion**
7	3	8	8×8	4	4×4	Octonion
8	2	4	4×4	2	2×2	Quaternion
9	1	2	2×2	0	1×1	Complex

Figure 6.1. The Octonion Cosmology ten space spectrum. The space for our universe, number 6, is in bold type.

The various columns in Fig. 6.1 are related through the equations:

$$q = 2n - 2 \qquad (6.1)$$

for the number of space-time dimensions and the Cayley number n. The Cayley number dimension d_c and the corresponding total size of the dimension array d_d are related by

$$d_c = 2^n \qquad (6.2)$$

$$d_d = 2^{2n} \qquad (6.3)$$

In addition

$$d_d^{(n)} = d_s^{(n+1)} \qquad (6.4)$$

---

[12] The space-time dimensions become 4 as in the Unified SuperStandard Theory through either transfer of dimensions to internal symmetries or by the compactification of two dimensions. As a result the pattern of fermion-antifermion annihilations producing spaces 7, 8, and 9 is disrupted.

where n is the Cayley-Dickson number of a space and n + 1 is the Cayley number of the space above it. The quantity $d_s^{(n+1)}$ is the total size of the $(n + 1)^{th}$ fermion spinor array.

    We also see the total number of spinor array components d_s for a space is

$$d_s = 2^{2n-2} \tag{6.5}$$
$$d_s = 2^r \tag{6.6}$$

implying

$$q = 2n - 2 \tag{6.7}$$

The spinor array's rows and columns sizes are

$$\text{Number of components in a spinor column} = 2^{n-1} \tag{6.8}$$

Consequently the fermion spin s is

$$s = (2^{n-2} - 1)/2 \tag{6.9}$$

These relations set the internal symmetry and space-time dimensions of the n > 4 spaces to those in Fig. 6.1.

6.1 Effect of Two Compactified Space 6 Dimensions

    The compactification of 6 dimension space-time to 4 dimensions (space number 6) distorts the spectrum in Fig. 6.1 since the spaces 7, 8, and 9 generated by fermion-antifermion annihilation are different. See Fig. 6.2 for a modified Octonion Spaces Spectrum.

Modified *TEN* OCTONION SPACES SPECTRUM

Spectrum Number

	Coordinate Cayley Type	Dimension of a Coordinates	Dimension Array Size d_d	Space-Time Dimension r
	Superverse Space			
0	Complex Octonion Octonion Octonion (1024)	Complex Octonion Octonion Octonion	1024×1024	18
1	Octonion Octonion Octonion (512)	Octonion Octonion Octonion	512×512	16
2	Quaternion Octonion Octonion (256)	Quaternion Octonion Octonion	256×256	14
3	Complex Octonion Octonion (128)	Complex Octonion Octonion	128×128	12
4	Octonion Octonion (64) Maxiverse Space	Octonion Octonion	64×64	10
5	Quaternion Octonion (32) Megaverses Space	Quaternion Octonion	32×32	8
6	Complex Octonion (16) Universe Space	Complex Octonion	16×16	4
	Minispaces			
7	Quaternion (4)	Quaternion	4×4	4
8	Real (4)	Real (4)	4×4	4
9	Real (4)	Real (4)	4×4	4

Figure 6.2. The spectrum of the ten octonion spaces. The spaces are numbered from 0 through 9. The numbers in parentheses in column 2 are the number of array row/column dimensions. The items in column 3 are the number of rows of dimensions (1024, 512, 256, 128, 64, 32, 16, 4, 4, 4). Spaces 7, 8, and 9 are for the case of fermion-antifermion annihilation in a 4 dimension universe (6 dimensions minus 2 compact dimensions).

7. Symmetry Splitting in Octonion Cosmology Based on Fermion-Antifermion Annihilation

This chapter shows the origin of symmetry splitting due to the chain of fermion-antifermion annihilations, and also due to the structure of the dimension arrays derived from spinor arrays.

7.1 General Pattern of Splittings due to Cayley Form of Spectrum

Cayley numbers are generated as powers of the integer two. Spinor Cayley number arrays are necessarily square. So octonion spectrum dimension arrays increase in size by factors of four. Thus the Cayley-Dickson number $n = 10$ space can be viewed as four copies of the $n = 9$ space. Their connection via fermion-antifermion annihilation indicates they are copies.

Iterating this process down the chain of annihilations to space creations leads to a composite view of dimension space 10 as a nested set of dimension arrays of the lower spaces. See Fig. 7.1.

The pattern of quadrisections gives the following partition of the $n = 10$ dimension array into dimension sub-blocks, some of which directly appear in QUeST and the New Unified SuperStandard Theory (NEWUST):

Array	Number of Components	QUeST Blocks
512 × 512	262,144	
128 × 128	16,384	
64 × 64	4096	
32 × 32	1024	size of Megaverse Array
16 × 16	256	size of Dimension Array
8 × 8	64	size of one layer
4 × 4	16	size of SU(2)⊗U(1)⊗SU(4)⊗SL(2, **C**) size of U(4)⊗U(4)
2 × 2	4	size of SU(2)⊗U(1) size of SL(2, **C**)

The symmetry structure of NEWQUeST and NEWUTMOST embody the structure of the chain of octonion spaces in chapter 5. The symmetry group structure of the octonion spaces is also directly connected to the structure of the spinor arrays within the scalar space particles created by fermion-antifermion annihilation seen earlier. The following section shows the connection in detail for the NEWUTMOST dimension array for the Megaverse.

The basis of the observed symmetries in the fermion-antifermion annihilation model of Octonion Cosmology raises the possibility that "symmetry breakdown" does not really apply to the observed symmetries except for symmetry breakdown in the ElectroWeak SU(2)⊗U(1) sectors.

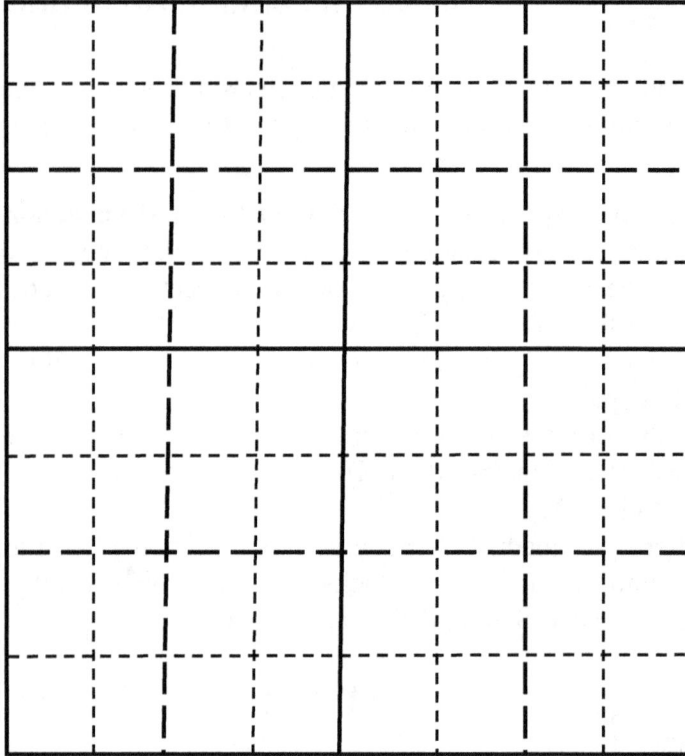

Figure 7.1. The n = 10 dimension array quadrisected three times into n = 7 dimension arrays. Quadrisection continues down to n = 1 for the spectrum of Fig. 6.1, and down to n = 4 (spectrum number 6) for the spectrum of Fig. 6.2.

7.2 Detailed Form of Splitting Based on Spinor Array Structure

The fermion-antifermion annihilation process was described in chapter 3. A fermion-antifermion pair with momenta p_+ and p_- generated a scalar space particle with momentum $p_s = p_+ + p_-$.

$$U(p_-) \ "\times" \ V(p_+) \to U(p_+ + p_-) \to DimensionArray \to Set\ of\ Symmetries$$

The spinor array of the created particle yielded a dimension array based on the $U(2^{r/2})$ symmetry. We now turn to consider the detailed form of the space-time 8 spinor array and show it reveals the symmetry of NEWQUeST and NEWUST.

The breakup of the spinor array into sets of large and small components is identical to the splitting of the dimension array into sets of dimensions of internal symmetries excepting ElectroWeak symmetry breakdowns.

7.2.0 Symmetry Splitting Based on Spinor Array Structure[13]

There is a general belief that there exists an overall symmetry for an entire universe/Megaverse. Spontaneous symmetry breakdowns are viewed as the source of the separation of particle symmetries into SU(3), SU(2)⊗U(1), and so on, as well as the breaking of symmetries such as ElectroWeak SU(2)⊗U(1) breakdown.

We put forward the proposition that the presumed inherent U(128) symmetry of 256 dimension QUeST universes (and similarly for Megaverses and other spaces) never existed.[14] Universes and Megaverses *began* with factored symmetries from their moment of origin in a fermion-antifermion annihilation.

We have described the splits generating the 10 octonion spaces, which we also believe existed from the first "moment" of existence of the Cosmos.

We see three types of symmetry splitting:

1. Splitting of the Superverse into 10 octonion spaces. We call this *global splitting by inheritance* since it generates entire octonion spaces.

2. Splitting of each octonion space into sets of Dimension-32 atoms. We call this type of splitting *local splitting by inheritance* since it splits the symmetry of an individual octonion space.

3. Splitting of a symmetry within a Dimension-32 atom such as SU(2)⊗U(1), SU(4), a U(4) Generation group, and a U(4) Layer group. This splitting is accomplished by spontaneous symmetry breaking.[15]

---

[13] This section previously appeared in Blaha (2021b).

[14] Chapter 6 suggests the Superverse is the origin of the ten octonion spaces. The ten spaces are shown as a "splitting" of the Superverse. As above, we propose that the split-generated factoring of spaces existed from the "Beginning" – not as a result of spontaneous symmetry breaking.

[15] At best, splitting of type 3 is the only splitting that may be relevant to running coupling constant estimates of the symmetry unification energy.

This section describes splitting of type 2. Spontaneous symmetry breaking (type 3) is described in Blaha (2018e) and (2020c).

7.2.1 Splitting of Type 2 Local Splitting by Inheritance

We saw in Blaha (2021a) that the overall structure of particle symmetries is determined in the large by the spinor structure of the annihilating fermion-antifermion pairs that generate universes[16] and megaverses.[17] *Thus there is a splitting of symmetries into a product of factors that is not due to spontaneous symmetry breaking (as it is usually envisioned.[18]) We call this type of splitting local splitting by inheritance since it is, in a very real sense, inherited from a "parent" octonion space instance.*

7.2.2 QUeST Inheritance

The spinor structure of an annihilating fermion-antifermion pair in an UTMOST Megaverse causes 4×4 blocks of dimensions in the resulting QUeST universe dimension array. It factors the overall symmetry into blocks of internal symmetries. The blocks are not the result of symmetry breaking but reflect the structure of the spinors in UTMOST Megaverse fermions.[19] The sixteen 16-spinors of an 8-dimension Megaverse fermion are depicted in Figs. 7.2 and 7.3.: Fig. 7.4 shows the resulting internal symmetry array 4×4 block structure of QUeST.

7.2.3 UTMOST Megaverse Inheritance

The spinor structure of an annihilating fermion-antifermion pair in the Maxiverse causes 8×8 blocks of dimensions in the resulting UTMOST Megaverse that determine blocks of internal symmetries.[20] These blocks (and the corresponding blocks for QUeST) are described in detail in chapter 8 of Blaha (2021a). The blocks are not the result of symmetry breaking but reflect the structure of the spinors in Maxiverse urfermions (fermions). The thirty-two 32-spinors of a 10 dimension Maxiverse urfermion are depicted in Fig. 7.5. Spinor parts map to 8×8 blocks of internal symmetries (Fig. 7.6) in the UTMOST Megaverse.

7.2.4 QUeST – UST Map and Map to UTMOST

The 4×4 blocks of QUeST are naturally determined by a map from QUeST to the Unified SuperStandard Theory (UST) that is shown in chapter 3 of Blaha (2019a). *The composition of the 8×8 blocks in the UTMOST Megaverse is determined by mapping upward from QUeST since the UTMOST dimension array consists of four copies of QUeST.*

---

[16] Chapter 4 of Blaha (2021a).

[17] Chapter 8 of Blaha (2021a).

[18] Spontaneous symmetry breaking does appear to take place in ElectroWeak Theory: SU(2)⊗U(1) breaking, and in the Generation and Layer groups, as well as the breaking of Strong groups from SU(4) to SU(3)⊗U(1).

[19] For that reason this form of symmetry factoring can be viewed as evidence for the existence of our universe in a Megaverse.

[20] The breakdown into 64 dimension blocks carries over to QUeST. Each layer of QUeST is a 64 dimension block.

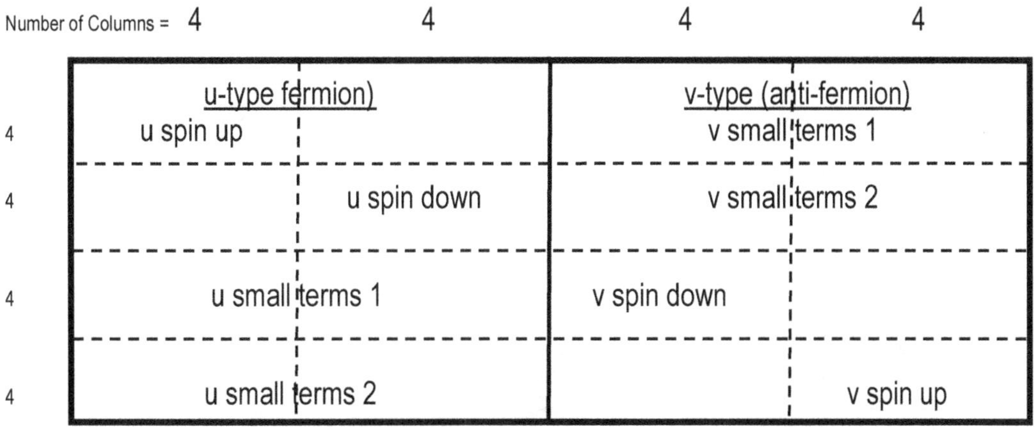
Number of Columns = 4 4 4 4
4
4
4
4
u-type fermion)
u spin up
u spin down
u small terms 1
u small terms 2
v-type (anti-fermion)
v small terms 1
v small terms 2
v spin down
v spin up

Figure 7.2. The 16-spinors of a 8 space-time dimension spinor. Each spinor column has 16 rows. There are 16 16-spinors.

u-up–v-v1	u-up–v-v2	u-up–v-down	u-up–v-up
u-up–v-v1	u-up–v-v2	u-up–v-down	u-up–v-up
u-up–v-v1	u-up–v-v2	u-up–v-down	u-up–v-up
u-up–v-v1	u-up–v-v2	u-up–v-down	u-up–v-up
u-down–v-v1	u-down–v-v2	u-down–v-down	u-down–v-up
u-down–v-v1	u-down–v-v2	u-down–v-down	u-down–v-up
u-down–v-v1	u-down–v-v2	u-down–v-down	u-down–v-up
u-down–v-v1	u-down–v-v2	u-down–v-down	u-down–v-up
u-v1–v-v1	u-v1–v-v2	u-v1–v-down	u-v1–v-up
u-v1–v-v1	u-v1–v-v2	u-v1–v-down	u-v1–v-up
u-v1–v-v1	u-v1–v-v2	u-v1–v-down	u-v1–v-up
u-v1–v-v1	u-v1–v-v2	u-v1–v-down	u-v1–v-up
u-v2–v-v1	u-v2–v-v2	u-v2–v-down	u-v2–v-up
u-v2–v-v1	u-v2–v-v2	u-v2–v-down	u-v2–v-up
u-v2–v-v1	u-v2–v-v2	u-v2–v-down	u-v2–v-up
u-v2–v-v1	u-v2–v-v2	u-v2–v-down	u-v2–v-up

Figure 7.3. The product array $[U_{ba}]$ of the composite u-type and v-type spinors illustrating the structure of the outer product array of uv's. The result of fermion-antifermion annihilation as indicated in chapter 3.

	4	4	4	4
4	u-up–v-v1	u-up–v-v2	u-up–v-down	u-up–v-up
4	u-down–v-v1	u-down–v-v2	u-down–v-down	u-down–v-up
4	u-v1–v-v1	u-v1–v-v2	u-v1–v-down	u-v1–v-up
4	u-v2–v-v1	u-v2–v-v2	u-v2–v-down	u-v2–v-up

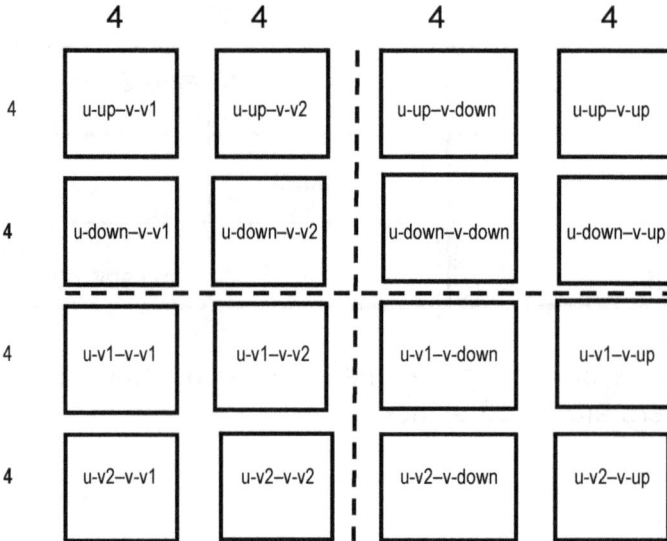

Figure 7.4. Block form of the 16 × 16 [U_{ba}] array. This is also the form of the QUeST dimension array of 256 dimensions. The blocks are divided by dashed lines that separate 64 dimension sections. These sections map to layers in QUeST (UST).

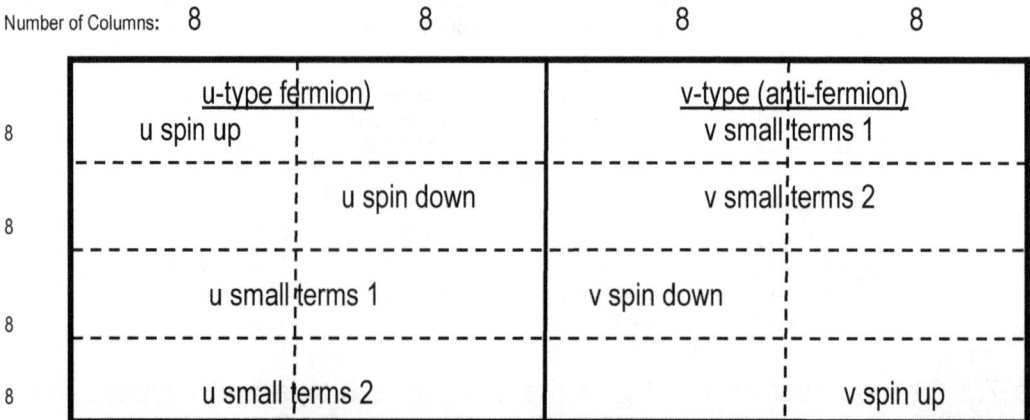

Number of Columns: 8   8   8   8

u-type fermion)		v-type (anti-fermion)	
u spin up		v small terms 1	
	u spin down	v small terms 2	
u small terms 1		v spin down	
u small terms 2			v spin up

Figure 7.5. The thirty-two 32-spinors of a 10 dimension urfermion (a Maxiverse fermion). Each spinor column has 32 rows.

u-up–v-v1	u-up–v-v2	u-up–v-down	u-up–v-up
u-up–v-v1	u-up–v-v2	u-up–v-down	u-up–v-up
u-up–v-v1	u-up–v-v2	u-up–v-down	u-up–v-up
u-up–v-v1	u-up–v-v2	u-up–v-down	u-up–v-up
u-up–v-v1	u-up–v-v2	u-up–v-down	u-up–v-up
u-up–v-v1	u-up–v-v2	u-up–v-down	u-up–v-up
u-up–v-v1	u-up–v-v2	u-up–v-down	u-up–v-up
u-up–v-v1	u-up–v-v2	u-up–v-down	u-up–v-up
u-down–v-v1	u-down–v-v2	u-down–v-down	u-down–v-up
u-down–v-v1	u-down–v-v2	u-down–v-down	u-down–v-up
u-down–v-v1	u-down–v-v2	u-down–v-down	u-down–v-up
u-down–v-v1	u-down–v-v2	u-down–v-down	u-down–v-up
u-down–v-v1	u-down–v-v2	u-down–v-down	u-down–v-up
u-down–v-v1	u-down–v-v2	u-down–v-down	u-down–v-up
u-down–v-v1	u-down–v-v2	u-down–v-down	u-down–v-up
u-down–v-v1	u-down–v-v2	u-down–v-down	u-down–v-up
u-v1–v-v1	u-v1–v-v2	u-v1–v-down	u-v1–v-up
u-v1–v-v1	u-v1–v-v2	u-v1–v-down	u-v1–v-up
u-v1–v-v1	u-v1–v-v2	u-v1–v-down	u-v1–v-up
u-v1–v-v1	u-v1–v-v2	u-v1–v-down	u-v1–v-up
u-v1–v-v1	u-v1–v-v2	u-v1–v-down	u-v1–v-up
u-v1–v-v1	u-v1–v-v2	u-v1–v-down	u-v1–v-up
u-v1–v-v1	u-v1–v-v2	u-v1–v-down	u-v1–v-up
u-v1–v-v1	u-v1–v-v2	u-v1–v-down	u-v1–v-up
u-v2–v-v1	u-v2–v-v2	u-v2–v-down	u-v2–v-up
u-v2–v-v1	u-v2–v-v2	u-v2–v-down	u-v2–v-up
u-v2–v-v1	u-v2–v-v2	u-v2–v-down	u-v2–v-up
u-v2–v-v1	u-v2–v-v2	u-v2–v-down	u-v2–v-up
u-v2–v-v1	u-v2–v-v2	u-v2–v-down	u-v2–v-up
u-v2–v-v1	u-v2–v-v2	u-v2–v-down	u-v2–v-up
u-v2–v-v1	u-v2–v-v2	u-v2–v-down	u-v2–v-up
u-v2–v-v1	u-v2–v-v2	u-v2–v-down	u-v2–v-up

Figure 7.6. The 32 × 32 product array of the composite u-type and v-type spinors illustrating the structure of the product array of uv's. Each column represents 8 spinor columns, making 32 columns in all. The result of fermion-antifermion annihilation as indicated in chapter 3.

8	8	8	8

8	u-up–v-v1	u-up–v-v2	u-up–v-down	u-up–v-up
8	u-down–v-v1	u-down–v-v2	u-down–v-down	u-down–v-up
8	u-v1–v-v1	u-v1–v-v2	u-v1–v-down	u-v1–v-up
8	u-v2–v-v1	u-v2–v-v2	u-v2–v-down	u-v2–v-up

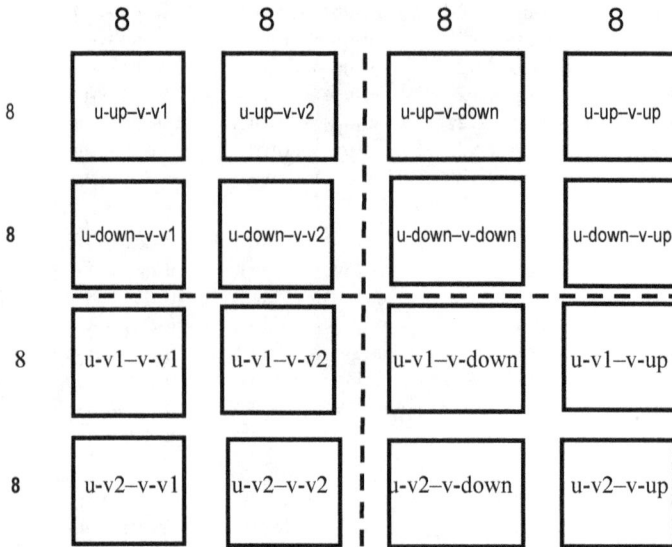

Figure 7.7. Block form of the resulting 32 × 32 array. This is also the form of the UTMOST dimension array of 1024 dimensions. It has 16 × 16 blocks, which have four 8 × 8 blocks within them. The set of four of 16 × 16 blocks above are divided by dashed lines into 256 dimension sections. These 256 dimension sections map to layers in UTMOST. Each UTMOST layer is equivalent to a QUeST 256 dimension array.

8. Dressing a Φ Space Particle

The Φ space particle defined in chapter 2 has a $U(2^{r/2})$ unitary group, which determines its internal symmetry and space-time groups.

However it does not appear to create universes, multiverses, and so on containing the usual accouterments of particles and energies of physical spaces. A dimension array, by itself, is not a universe, multiverse, and so on. We must "dress" each space with particles and energy.

The situation is not unlike C++ where one defines a class specifying variables within it, and then creates objects of the class type by allocating chunks of memory.

We turn now to the analogous dressing of spaces with symmetries, energy and particles.

8.1 Symmetry Dressing

The dimension array resulting from a spinor array created from a fermion-antifermion annihilation into a scalar particle enables a set of symmetries to be allocated by splitting dimensions into sets of fundamental representation dimensions. This conceptual process requires no work. But it creates a framework for the specification of the elementary particle spectrum.

8.2 Energy/Particle Dressing

Each fundamental symmetry has a set of fundamental fermions in a fundamental representation. Each symmetry also has associated sets of gauge vector bosons and Higgs bosons. To those particles we add, at minimum, Gravitation and its gravitons.

The generation of the particles requires energy to be present. The mass m' of the created scalar boson is the likely source of the energy of the created instance of the space. Thus the source of the energy of a child space instance is in the parent space instance.[21]

It is reasonable to assume that the mass-energy is initially concentrated at the creation point. It is also reasonable to assume the energy embodied in the particles has a Planck black body distribution.[22] The expansion of the space follows.

8.3 A View of Creation

We have seen the progression from a new Quantum Field Theory of Spaces to Octonion Cosmology with a fairly detailed relation between the spinor array generated by an annihilating fermion-antifermion pair and space internal symmetries. This achievement raises the possibility that quantum field theory, which originated in the time of electrons, protons and neutrons, may be the true "language" of Nature.[23] It

---

[21] The energy of the top space (God Space) is assumed. All energies of all child spaces flow from it.

[22] Blaha (2004) and (2021d).

[23] The similarity of QUeST structure and DNA structure also raises the possibility of a stability principle for structures that applies to both DNA and Octonion Cosmology. See Blaha (2021e).

appears that a more fundamental substratum of Reality exists, from which Quantum Field Theory proceeds.

REFERENCES

Akhiezer, N. I., Frink, A. H. (tr), 1962, *The Calculus of Variations* (Blaisdell Publishing, New York, 1962).

Bjorken, J. D., Drell, S. D., 1964, *Relativistic Quantum Mechanics* (McGraw-Hill, New York, 1965).

Bjorken, J. D., Drell, S. D., 1965, *Relativistic Quantum Fields* (McGraw-Hill, New York, 1965).

Blaha, S., 1995, *C++ for Professional Programming* (International Thomson Publishing, Boston, 1995).

_____, 1998, *Cosmos and Consciousness* (Pingree-Hill Publishing, Auburn, NH, 1998 and 2002).

_____, 2002, *A Finite Unified Quantum Field Theory of the Elementary Particle Standard Model and Quantum Gravity Based on New Quantum Dimensions™ & a New Paradigm in the Calculus of Variations* (Pingree-Hill Publishing, Auburn, NH, 2002).

_____, 2004, *Quantum Big Bang Cosmology: Complex Space-time General Relativity, Quantum Coordinates,™ Dodecahedral Universe, Inflation, and New Spin 0, ½, 1 & 2 Tachyons & Imagyons* (Pingree-Hill Publishing, Auburn, NH, 2004).

_____, 2005a, *Quantum Theory of the Third Kind: A New Type of Divergence-free Quantum Field Theory Supporting a Unified Standard Model of Elementary Particles and Quantum Gravity based on a New Method in the Calculus of Variations* (Pingree-Hill Publishing, Auburn, NH, 2005).

_____, 2005b, *The Metatheory of Physics Theories, and the Theory of Everything as a Quantum Computer Language* (Pingree-Hill Publishing, Auburn, NH, 2005).

_____, 2005c, *The Equivalence of Elementary Particle Theories and Computer Languages: Quantum Computers, Turing Machines, Standard Model, Superstring Theory, and a Proof that Gödel's Theorem Implies Nature Must Be Quantum* (Pingree-Hill Publishing, Auburn, NH, 2005).

_____, 2006a, *The Foundation of the Forces of Nature* (Pingree-Hill Publishing, Auburn, NH, 2006).

_____, 2006b, *A Derivation of ElectroWeak Theory based on an Extension of Special Relativity; Black Hole Tachyons; & Tachyons of Any Spin.* (Pingree-Hill Publishing, Auburn, NH, 2006).

_____, 2007a, *Physics Beyond the Light Barrier: The Source of Parity Violation, Tachyons, and A Derivation of Standard Model Features* (Pingree-Hill Publishing, Auburn, NH, 2007).

_____, 2007b, *The Origin of the Standard Model: The Genesis of Four Quark and Lepton Species, Parity Violation, the ElectroWeak Sector, Color SU(3), Three Visible Generations of Fermions, and One Generation of Dark Matter with Dark Energy* (Pingree-Hill Publishing, Auburn, NH, 2007).

_____, 2008a, *A Direct Derivation of the Form of the Standard Model From GL(16)* (Pingree-Hill Publishing, Auburn, NH, 2008).

_____, 2008b, *A Complete Derivation of the Form of the Standard Model With a New Method to Generate Particle Masses Second Edition* (Pingree-Hill Publishing, Auburn, NH, 2008)

_____, 2009, *The Algebra of Thought & Reality: The Mathematical Basis for Plato's Theory of Ideas, and Reality Extended to Include A Priori Observers and Space-Time Second Edition* (Pingree-Hill Publishing, Auburn, NH, 2009).

_____, 2010a, *Operator Metaphysics: A New Metaphysics Based on a New Operator Logic and a New Quantum Operator Logic that Lead to a Mathematical Basis for Plato's Theory of Ideas and Reality* (Pingree-Hill Publishing, Auburn, NH, 2010).

_____, 2010b, *The Standard Model's Form Derived from Operator Logic, Superluminal Transformations and GL(16)* (Pingree-Hill Publishing, Auburn, NH, 2010).

_____, 2010c, *SuperCivilizations: Civilizations as Superorganisms* (McMann-Fisher Publishing, Auburn, NH, 2010).

_____, 2011a, *21st Century Natural Philosophy Of Ultimate Physical Reality* (McMann-Fisher Publishing, Auburn, NH, 2011).

_____, 2011b, *All the Universe! Faster Than Light Tachyon Quark Starships & Particle Accelerators with the LHC as a Prototype Starship Drive Scientific Edition* (Pingree-Hill Publishing, Auburn, NH, 2011).

_____, 2011c, *From Asynchronous Logic to The Standard Model to Superflight to the Stars* (Blaha Research, Auburn, NH, 2011).

_____, 2012a, *From Asynchronous Logic to The Standard Model to Superflight to the Stars volume 2: Superluminal CP and CPT, U(4) Complex General Relativity and The Standard Model, Complex Vierbein General Relativity, Kinetic Theory, Thermodynamics* (Blaha Research, Auburn, NH, 2012).

_____, 2012b, *Standard Model Symmetries, And Four And Sixteen Dimension Complex Relativity; The Origin Of Higgs Mass Terms* (Blaha Reasearch, Auburn, NH, 2012).

_____, 2013a, *Multi-Stage Space Guns, Micro-Pulse Nuclear Rockets, and Faster-Than-Light Quark-Gluon Ion Drive Starships* (Blaha Research, Auburn, NH, 2013).

_____, 2013b, *The Bridge to Dark Matter; A New Sister Universe; Dark Energy; Inflatons; Quantum Big Bang; Superluminal Physics; An Extended Standard Model Based on Geometry* (Blaha Reasearch, Auburn, NH, 2013).

_____, 2014a, *Universes and Megaverses: From a New Standard Model to a Physical Megaverse; The Big Bang; Our Sister Universe's Wormhole; Origin of the Cosmological Constant, Spatial Asymmetry of the Universe, and its Web of Galaxies; A Baryonic Field between Universes and Particles; Megaverse Extended Wheeler-DeWitt Equation* (Blaha Reasearch, Auburn, NH, 2014).

_____, 2014b, *All the Megaverse! Starships Exploring the Endless Universes of the Cosmos Using the Baryonic Force* (Blaha Research, Auburn, NH, 2014).

_____, 2014c, *All the Megaverse! II Between Megaverse Universes: Quantum Entanglement Explained by the Megaverse Coherent Baryonic Radiation Devices – PHASERs Neutron Star Megaverse Slingshot*

Dynamics Spiritual and UFO Events, and the Megaverse Microscopic Entry into the Megaverse (Blaha Research, Auburn, NH, 2014).

_____, 2015a, *PHYSICS IS LOGIC PAINTED ON THE VOID: Origin of Bare Masses and The Standard Model in Logic, U(4) Origin of the Generations, Normal and Dark Baryonic Forces, Dark Matter, Dark Energy, The Big Bang, Complex General Relativity, A Megaverse of Universe Particles* (Blaha Research, Auburn, NH, 2015).

_____, 2015b, *PHYSICS IS LOGIC Part II: The Theory of Everything, The Megaverse Theory of Everything, U(4)$\otimes$U(4) Grand Unified Theory (GUT), Inertial Mass = Gravitational Mass, Unified Extended Standard Model and a New Complex General Relativity with Higgs Particles, Generation Group Higgs Particles* (Blaha Research, Auburn, NH, 2015).

_____, 2015c, *The Origin of Higgs ("God") Particles and the Higgs Mechanism: Physics is Logic III, Beyond Higgs – A Revamped Theory With a Local Arrow of Time, The Theory of Everything Enhanced, Why Inertial Frames are Special, Universes of the Mind* (Blaha Research, Auburn, NH, 2015).

_____, 2015d, *The Origin of the Eight Coupling Constants of The Theory of Everything: U(8) Grand Unified Theory of Everything (GUTE), S^8 Coupling Constant Symmetry, Space-Time Dependent Coupling Constants, Big Bang Vacuum Coupling Constants, Physics is Logic IV* (Blaha Research, Auburn, NH, 2015).

_____, 2016a, *New Types of Dark Matter, Big Bang Equipartition, and A New U(4) Symmetry in the Theory of Everything: Equipartition Principle for Fermions, Matter is 83.33% Dark, Penetrating the Veil of the Big Bang, Explicit QFT Quark Confinement and Charmonium, Physics is Logic V* (Blaha Research, Auburn, NH, 2016).

_____, 2016b, *The Periodic Table of the 192 Quarks and Leptons in The Theory of Everything: The U(4) Layer Group, Physics is Logic VI* (Blaha Research, Auburn, NH, 2016).

_____, 2016c, *New Boson Quantum Field Theory, Dark Matter Dynamics, Dark Matter Fermion Layer Mixing, Genesis of Higgs Particles, New Layer Higgs Masses, Higgs Coupling Constants, Non-Abelian Higgs Gauge Fields, Physics is Logic VII* (Blaha Research, Auburn, NH, 2016).

_____, 2016d, *Unification of the Strong Interactions and Gravitation: Quark Confinement Linked to Modified Short-Distance Gravity; Physics is Logic VIII* (Blaha Research, Auburn, NH, 2016).

_____, 2016e, *MoND: Unification of the Strong Interactions and Gravitation II, Quark Confinement Linked to Large-Scale Gravity, Physics is Logic IX* (Blaha Research, Auburn, NH, 2016).

_____, 2016f, *CQ Mechanics: A Unification of Quantum & Classical Mechanics, Quantum/Semi-Classical Entanglement, Quantum/Classical Path Integrals, Quantum/Classical Chaos* (Blaha Research, Auburn, NH, 2016).

_____, 2016g, *GEMS: Unified Gravity, ElectroMagnetic and Strong Interactions: Manifest Quark Confinement, A Solution for the Proton Spin Puzzle, Modified Gravity on the Galactic Scale* (Pingree Hill Publishing, Auburn, NH, 2016).

_____, 2016h, *Unification of the Seven Boson Interactions based on the Riemann-Christoffel Curvature Tensor* (Pingree Hill Publishing, Auburn, NH, 2016).

_____, 2017a, *Unification of the Eleven Boson Interactions based on 'Rotations of Interactions'* (Pingree Hill Publishing, Auburn, NH, 2017).

_____, 2017b, *The Origin of Fermions and Bosons, and Their Unification* (Pingree Hill Publishing, Auburn, NH, 2017).

_____, 2017c, *Megaverse: The Universe of Universes* (Pingree Hill Publishing, Auburn, NH, 2017).

_____, 2017d, *SuperSymmetry and the Unified SuperStandard Model* (Pingree Hill Publishing, Auburn, NH, 2017).

_____, 2017e, *From Qubits to the Unified SuperStandard Model with Embedded SuperStrings: A Derivation* (Pingree Hill Publishing, Auburn, NH, 2017).

_____, 2017f, *The Unified SuperStandard Model in Our Universe and the Megaverse: Quarks, ... ,* (Pingree Hill Publishing, Auburn, NH, 2017).

_____, 2018a, *The Unified SuperStandard Model and the Megaverse SECOND EDITION A Deeper Theory based on a New Particle Functional Space that Explicates Quantum Entanglement Spookiness (Volume 1)* (Pingree Hill Publishing, Auburn, NH, 2018).

_____, 2018b, *Cosmos Creation: The Unified SuperStandard Model, Volume 2, SECOND EDITION* (Pingree Hill Publishing, Auburn, NH, 2018).

_____, 2018c, *God Theory (*Pingree Hill Publishing, Auburn, NH, 2018).

_____, 2018d, *Immortal Eye: God Theory: Second Edition* (Pingree Hill Publishing, Auburn, NH, 2018).

_____, 2018e, *Unification of God Theory and Unified SuperStandard Model THIRD EDITION* (Pingree Hill Publishing, Auburn, NH, 2018).

_____, 2019a, *Calculation of: QED α = 1/137, and Other Coupling Constants of the Unified SuperStandard Theory* (Pingree Hill Publishing, Auburn, NH, 2019).

_____, 2019b, *Coupling Constants of the Unified SuperStandard Theory SECOND EDITION* (Pingree Hill Publishing, Auburn, NH, 2019).

_____, 2019c, *New Hybrid Quantum Big_Bang–Megaverse_Driven Universe with a Finite Big Bang and an Increasing Hubble Constant* (Pingree Hill Publishing, Auburn, NH, 2019).

_____, 2019d, *The Universe, The Electron and The Vacuum* (Pingree Hill Publishing, Auburn, NH, 2019).

_____, 2019e, *Quantum Big Bang – Quantum Vacuum Universes (Particles)* (Pingree Hill Publishing, Auburn, NH, 2019).

_____, 2019f, *The Exact QED Calculation of the Fine Structure Constant Implies ALL 4D Universes have the Same Physics/Life Prospects* (Pingree Hill Publishing, Auburn, NH, 2019).

_____, 2019g, *Unified SuperStandard Theory and the SuperUniverse Model: The Foundation of Science* (Pingree Hill Publishing, Auburn, NH, 2019).

_____, 2020a, *Quaternion Unified SuperStandard Theory (The QUeST) and Megaverse Octonion SuperStandard Theory (MOST)* (Pingree Hill Publishing, Auburn, NH, 2020).

_____, 2020b, *United Universes Quaternion Universe - Octonion Megaverse* (Pingree Hill Publishing, Auburn, NH, 2020).

_____, 2020c, *Unified SuperStandard Theories for Quaternion Universes & The Octonion Megaverse* (Pingree Hill Publishing, Auburn, NH, 2020).

_____, 2020d, *The Essence of Eternity: Quaternion & Octonion SuperStandard Theories* (Pingree Hill Publishing, Auburn, NH, 2020).

_____, 2020e, *The Essence of Eternity II* (Pingree Hill Publishing, Auburn, NH, 2020).

_____, 2020f, *A Very Conscious Universe* (Pingree Hill Publishing, Auburn, NH, 2020).

_____, 2020g, *Hypercomplex Universe* (Pingree Hill Publishing, Auburn, NH, 2020).

_____, 2020h, *Beneath the Quaternion Universe* (Pingree Hill Publishing, Auburn, NH, 2020).

_____, 2020i, *Why is the Universe Real? From Quaternion & Octonion to Real Coordinates* (Pingree Hill Publishing, Auburn, NH, 2020).

_____, 2020j, *The Origin of Universes: of Quaternion Unified SuperStandard Theory (QUeST); and of the Octonion Megaverse (UTMOST)* (Pingree Hill Publishing, Auburn, NH, 2020).

_____, 2020k, *The Seven Spaces of Creation: Octonion Cosmology* (Pingree Hill Publishing, Auburn, NH, 2020).

_____, 2020l, *From Octonion Cosmology to the Unified SuperStandard Theory of Particles* (Pingree Hill Publishing, Auburn, NH, 2020).

_____, 2021a, *Pioneering the Cosmos* (Pingree Hill Publishing, Auburn, NH, 2021).

_____, 2021b, *Pioneering the Cosmos II* (Pingree Hill Publishing, Auburn, NH, 2021).

_____, 2021c, *Beyond Octonion Cosmology* (Pingree Hill Publishing, Auburn, NH, 2021).

_____, 2021d, *Universes are Particles* (Pingree Hill Publishing, Auburn, NH, 2021).

_____, 2021e, *Octonion-like dna-based life, Universe expansion is decay, Emerging New Physics* (Pingree Hill Publishing, Auburn, NH, 2021).

Eddington, A. S., 1952, *The Mathematical Theory of Relativity* (Cambridge University Press, Cambridge, U.K., 1952).

Fant, Karl M., 2005, *Logically Determined Design: Clockless System Design With NULL Convention Logic* (John Wiley and Sons, Hoboken, NJ, 2005).

Feinberg, G. and Shapiro, R., 1980, *Life Beyond Earth: The Intelligent Earthlings Guide to Life in the Universe* (William Morrow and Company, New York, 1980).

Gelfand, I. M., Fomin, S. V., Silverman, R. A. (tr), 2000, *Calculus of Variations* (Dover Publications, Mineola, NY, 2000).

Giaquinta, M., Modica, G., Souchek, J., 1998, *Cartesian Coordinates in the Calculus of Variations* Volumes I and II (Springer-Verlag, New York, 1998).

Giaquinta, M., Hildebrandt, S., 1996, *Calculus of Variations* Volumes I and II (Springer-Verlag, New York, 1996).

Gradshteyn, I. S. and Ryzhik, I. M., 1965, *Table of Integrals, Series, and Products* (Academic Press, New York, 1965).

Heitler, W., 1954, *The Quantum Theory of Radiation* (Claendon Press, Oxford, UK, 1954).

Huang, Kerson, 1992, *Quarks, Leptons & Gauge Fields 2nd Edition* (World Scientific Publishing Company, Singapore, 1992).

Jost, J., Li-Jost, X., 1998, *Calculus of Variations* (Cambridge University Press, New York, 1998).

Kaku, Michio, 1993, *Quantum Field Theory*, (Oxford University Press, New York, 1993).

Kirk, G. S. and Raven, J. E., 1962, *The Presocratic Philosophers* (Cambridge University Press, New York, 1962).

Landau, L. D. and Lifshitz, E. M., 1987, *Fluid Mechanics 2nd Edition*, (Pergamon Press, Elmsford, NY, 1987).

Misner, C. W., Thorne, K. S., and Wheeler, J. A., 1973, *Gravitation* (W. H. Freeman, New York, 1973).

Rescher, N., 1967, *The Philosophy of Leibniz* (Prentice-Hall, Englewood Cliffs, NJ, 1967).

Rieffel, Eleanor and Polak, Wolfgang, 2014, *Quantum Computing* (MIT Press, Cambridge, MA, 2014).

Riesz, Frigyes and Sz.-Nagy, Béla, 1990, *Functional Analysis* (Dover Publications, New York, 1990).

Sagan, H., 1993, *Introduction to the Calculus of Variations* (Dover Publications, Mineola, NY, 1993).

Sakurai, J. J., 1964, *Invariance Principles and Elementary Particles* (Princeton University Press, Princeton, NJ, 1964).

Weinberg, S., 1972, *Gravitation and Cosmology* (John Wiley and Sons, New York, 1972).

Weinberg, S., 1995, *The Quantum Theory of Fields Volume I* (Cambridge University Press, New York, 1995).

INDEX

About the Author

Stephen Blaha is a well-known Physicist and Man of Letters with interests in Science, Society and civilization, the Arts, and Technology. He had an Alfred P. Sloan Foundation scholarship in college. He received his Ph.D. in Physics from Rockefeller University. He has served on the faculties of several major universities. He was also a Member of the Technical Staff at Bell Laboratories, a manager at the Boston Globe Newspaper, a Director at Wang Laboratories, and President of Blaha Software Inc. and of Janus Associates Inc. (NH).

Among other achievements he was a co-discoverer of the "r potential" for heavy quark binding developing the first (and still the only demonstrable) non-Aeolian gauge theory with an "r" potential; first suggested the existence of topological structures in superfluid He-3; first proposed Yang-Mills theories would appear in condensed matter phenomena with non-scalar order parameters; first developed a grammar-based formalism for quantum computers and applied it to elementary particle theories; first developed a new form of quantum field theory without divergences (thus solving a major 60 year old problem that enabled a unified theory of the Standard Model and Quantum Gravity without divergences to be developed); first developed a formulation of complex General Relativity based on analytic continuation from real space-time; first developed a generalized non-homogeneous Robertson-Walker metric that enabled a quantum theory of the Big Bang to be developed without singularities at t = 0; first generalized Cauchy's theorem and Gauss' theorem to complex, curved multi-dimensional spaces; received Honorable Mention in the Gravity Research Foundation Essay Competition in 1978; first developed a physically acceptable theory of faster-than-light particles; first derived a composition of extremums method in the Calculus of Variations; first quantitatively suggested that inflationary periods in the history of the universe were not needed; first proved Gödel's Theorem implies Nature must be quantum; provided a new alternative to the Higgs Mechanism, and Higgs particles, to generate masses; first showed how to resolve logical paradoxes including Gödel's Undecidability Theorem by developing Operator Logic and Quantum Operator Logic; first developed a quantitative harmonic oscillator-like model of the life cycle, and interactions, of civilizations; first showed how equations describing superorganisms also apply to civilizations. A recent book shows his theory applies successfully to the past 14 years of history and to *new* archaeological data on Andean and Mayan civilizations as well as Early Anatolian and Egyptian civilizations.

He first developed an axiomatic derivation of the form of The Standard Model from geometry – space-time properties – The Unified SuperStandard Model. It unifies all the known forces of Nature. It also has a Dark Matter sector that includes a Dark ElectroWeak sector with Dark doublets and Dark gauge interactions. It uses quantum coordinates to remove infinities that crop up in most

interacting quantum field theories and additionally to remove the infinities that appear in the Big Bang and generate inflationary growth of the universe. It shows gravity has a MOND-like form without sacrificing Newton's Laws. It relates the interactions of the MOND-like sector of gravity with the r-potential of Quark Confinement. The axioms of the theory lead to the question of their origin. We suggest in the preceding edition of this book it can be attributed to an entity with God-like properties. We explore these properties in "God Theory" and show they predict that the Cosmos exists forever although individual universes (or incarnations of our universe) "come and go." Several other important results emerge from God Theory such a functionally triune God. The Unified SuperStandard Theory has many other important parts described in the Current Edition of *The Unified SuperStandard Theory* and expanded in subsequent volumes.

Blaha has had a major impact on a succession of elementary particle theories: his Ph.D. thesis (1970), and papers, showed that quantum field theory calculations to all orders in ladder approximations could not give scaling deep inelastic electron-nucleon scattering. He later showed the eigenvalue equation for the fine structure constant α in Johnson-Baker-Willey QED had a zero at $\alpha = 1$ not 1/137 by solving the Schwinger-Dyson equations to all orders in an approximation that agreed with exact results to 4th order in α thus ending interest in this theory. In 1979 at Prof. Ken Johnson's (MIT) suggestion he calculated the proton-neutron mass difference in the MIT bag model and found the result had the wrong sign reducing interest in the bag model. These results all appear in Physical Review papers. In the 2000's he repeatedly pointed out the shortcomings of SuperString theory and showed that The Standard Model's form could be derived from space-time geometry by an extension of Lorentz transformations to faster than light transformations. This deeper space-time basis greatly increases the possibility that it is part of THE fundamental theory. Recently, Blaha showed that the Weak interactions differed significantly from the Strong, electromagnetic and gravitation interactions in important respects while these interactions had similar features, and suggested that ElectroWeak theory, which is essentially a glued union of the Weak interactions and Electromagnetism, possibly modulo unknown Higgs particle features, be replaced by a unified theory of the other interactions combined with a stand-alone Weak interaction theory. Blaha also showed that, if Charmonium calculations are taken seriously, the Strong interaction coupling constant is only a factor of five larger than the electromagnetic coupling constant, and thus Strong interaction perturbation theory would make sense and yield physically meaningful results.

In graduate school (1965-71) he wrote substantial papers in elementary particles and group theory: The Inelastic E- P Structure Functions in a Gluon Model. Phys. Lett. B40:501-502,1972; Deep-Inelastic E-P Structure Functions In A Ladder Model With Spin 1/2 Nucleons, Phys.Rev. D3:510-523,1971; Continuum Contributions To The Pion Radius, Phys. Rev. 178:2167-2169,1969; Character Analysis of U(N) and SU(N), J. Math. Phys. <u>10</u>, 2156 (1969); and The Calculation of the Irreducible Characters of the Symmetric Group in Terms of the

Compound Characters, (Published as Blaha's Lemma in D. E. Knuth's book: *The Art of Computer Programming Vols. 1 – 4*).

In the early 1980's Blaha was also a pioneer in the development of UNIX for financial, scientific and Internet applications: benchmarked UNIX versions showing that block size was critical for UNIX performance, developing financial modeling software, starting database benchmarking comparison studies, developing Internet-like UNIX networking (1982) and developing a hybrid shell programming technique (1982) that was a precursor to the PERL programming language. He was also the manager of the AT&T ten-year future products development database. His work helped lead to commercial UNIX on computers such as Sun Micros, IBM AIX minis, and Apple computers.

In the 1980's he pioneered the development of PC Desktop Publishing on laser printers and was nominated for three "Awards for Technical Excellence" in 1987 by PC Magazine for PC software products that he designed and developed.

Recently he has developed a theory of Megaverses – actual universes of which our universe is one – with quantum particle-like properties based on the Wheeler-DeWitt equation of Quantum Gravity. He has developed a theory of a baryonic force, which had been conjectured many years ago, and estimated the strength of the force based on discrepancies in measurements of the gravitational constant G. This force, operative in D-dimensional space, can be used to escape from our universe in "uniships" which are the equivalent of the faster-than-light starships proposed in the author's earlier books. Thus travel to other universes, as well as to other stars is possible.

Blaha also considered the complexified Wheeler-DeWitt equation and showed that its limitation to real-valued coordinates and metrics generated a Cosmological Constant in the Einstein equations.

The author has also recently written a series of books on the serious problems of the United States and their solution as well as a book on the decline of Mankind that will follow from current social and genetic trends in Mankind.

In the past twenty years Dr. Blaha has written over 80 books on a wide range of topics. Some recent major works are: *From Asynchronous Logic to The Standard Model to Superflight to the Stars, All the Universe!, SuperCivilizations: Civilizations as Superorganisms, America's Future: an Islamic Surge, ISIS, al Qaeda, World Epidemics, Ukraine, Russia-China Pact, US Leadership Crisis, The Rises and Falls of Man – Destiny – 3000 AD: New Support for a Superorganism MACRO-THEORY of CIVILIZATIONS From CURRENT WORLD TRENDS and NEW Peruvian, Pre-Mayan, Mayan, Anatolian, and Early Egyptian Data, with a Projection to 3000 AD*, and *Mankind in Decline: Genetic Disasters, Human-Animal Hybrids, Overpopulation, Pollution, Global Warming, Food and Water Shortages, Desertification, Poverty, Rising Violence, Genocide, Epidemics, Wars, Leadership Failure.*

He has taught approximately 4,000 students in undergraduate, graduate, and postgraduate corporate education courses primarily in major universities, and large companies and government agencies.

Recently he developed a quantum theory, The Unified SuperStandard Theory (UST), which describes elementary particles in detail without the difficulties of conventional quantum field theory. He found that the internal symmetries of this theory could be exactly derived from an octonion theory called QUeST. He further found that another octonion theory (UTMOST) describes the Megaverse. It can hold QUeST universes such as our own universe. It has an internal symmetry structure which is a superset of the QUeST internal symmetries. This book provides a complete derivation from the Quantum Field Theory of Spaces.